Pico-solar Electric Systems

The Earthscan Expert Guide to the Technology and Emerging Market

John Keane

Routledge
Taylor & Francis Group
LONDON AND NEW YORK

earthscan
from Routledge

First edition published 2014
by Routledge
2 Park Square, Milton Park, Abingdon, Oxon OX14 4RN

Simultaneously published in the USA and Canada
by Routledge
711 Third Avenue, New York, NY 10017

Routledge is an imprint of the Taylor & Francis Group, an informa business

British Library Cataloguing in Publication Data
A catalogue record for this book is available from the British Library

Library of Congress Cataloging in Publication Data
Keane, John (Urban planner)
Pico-solar electric systems : the Earthscan expert guide to the Technology and Emerging Market / John Keane. — First edition.
pages cm — (Earthscan expert series)
Includes bibliographical references and index.
1. Building-integrated photovoltaic systems. 2. Small power production facilities.
3. Solar houses. I. Title.
TK1087.K43 2014
621.31′244—dc23 2013035359

ISBN: 978-0-415-82359-3 (hbk)
ISBN: 978-1-315-81846-7 (ebk)

Typeset in Sabon
by Keystroke, Station Road, Codsall, Wolverhampton
Printed in Great Britain by Ashford Colour Press Ltd

MIX
Paper from responsible sources
FSC® C011748
www.fsc.org

Contents

Illustrations

Figures

Tables

Preface

Over 1.6 billion people across the world currently live without access to electricity. This means that around one-quarter of the world's population are forced to rely on outdated, expensive, poor quality and often dangerous fuels such as kerosene, candles and disposable batteries to meet many of their basic energy needs and avoid sitting in darkness each night. The development sector is forever setting new targets and initiatives aimed at reducing poverty, while neglecting to address this basic human need. Meanwhile, it is the mobile phone which has arguably had one of the greatest impacts on life across the planet in recent times.

Just as the mobile phone effectively enabled people to leapfrog the need to connect to a landline, (many people would still be waiting today for landline connections), a new category of low power, solar electric systems: 'pico-solar systems', offer the opportunity for people to access small, but incredibly useful amounts of clean, renewable electricity to transform their lives, wherever they live or are travelling to. This means that tens of millions of people no longer have to wait for an electricity grid which may never arrive just to turn on electric lights and charge up phones as well as the ever increasing range of hi-tech, low power consumption, appliances which exist in today's world.

This book provides a comprehensive overview of the pico-solar sector, from the technology behind the pico-solar revolution to how systems are transforming the lives of millions of people who live without access to electricity. As the largest potential market for pico-solar systems is across rural Africa and Asia, this book focuses on the challenges the sector faces in developing these markets.

Pico-solar systems can also offer useful amounts of power to a range of alternative customers, from festival-goers and travellers who want to keep their phone charged, to local authorities looking for more environmentally friendly ways in which to provide power to city parking meters. This book includes examples, and where necessary offers a critique of the increasing variety of applications pico-solar can be used for.

On a personal note, I experienced what life is like without access to electricity after living in a rural village in Tanzania, called Uhomini, in 2000. As I walked along the dirt road away from the village, with Ellen, a fellow volunteer, I waved goodbye to a small four-year-old boy called Festo, who had visited our house every day and probably could not believe that we were leaving. It was then I realised that the village was not going to change anytime soon. It would probably be many decades before it would benefit from basic amenities like electricity and running water in every house. Over a decade later, that village is still in the dark each night, waiting for an electricity grid that may never come. Festo is now 18 years old.

The true power of pico-solar is that it can bring electric light and much more to Uhomini and the millions of villages like it across the world. It can do this today, without any more waiting.

What This Book is About

This book introduces pico-solar electric systems, a new and rapidly expanding category of solar photovoltaic systems designed to produce small, but very useful amounts of power to charge an ever increasing array of low energy appliances. Today, pico-solar systems are providing power to millions of people and transforming the lives of off-grid households across rural Asia and Africa. Pico-solar products and systems are being used to power an increasingly wide range of appliances, from energy efficient LED lighting, to mobile phones, cameras, radios, MP3 players, e-readers and even parking meters in high streets.

This book is for social and environmentally driven people interested in learning how small amounts of renewable energy can make a big difference to the world we live in. It will be of particular interest to students, entrepreneurs, development actors and solar manufacturers and anyone who wishes:

- to learn more about pico-solar technology and how it differs from traditional solar home systems;
- advice on how to choose a quality pico-solar system and ensure it is kept in good working order;
- to understand the type of appliance pico-solar systems can charge power and those which require more electricity to operate than pico-solar systems can provide;
- to understand the positive, often transformative, impact systems can have, particularly on the lives of people who live without access to electricity;
- to understand the challenges which must be overcome in order to build a sustainable market and learn from leading examples of innovative companies and initiatives from across the world.

This book covers a wide range of topics.

Chapter 1 introduces a range of pico-solar product examples and provides an overview of the pico-solar markets across the world.

Chapter 2 explains basic solar principles and how they relate to pico-solar electric systems.

Chapter 3 summarises how solar cells and modules work and discusses the different types of photovoltaic (PV) technologies.

Chapter 4 explains how batteries work and introduces the range of battery chemistries which are used in pico-solar systems.

Chapter 5 explains basic lighting principles, measurements and provides an overview of LED lighting technology.

Chapter 6 explains how to calculate energy needs, understand system sizes and provides examples of the increasing range of appliances which can be powered. This chapter also provides examples of solar systems which generate and store more electricity than typical pico-solar systems, but are similar in every other respect.

Chapter 7 provides an overview of the international and industry standards designed to protect the consumer from poor quality and underperforming products, and provides guidance on how to conduct simple product tests outside of a laboratory.

Chapter 8 explains how to maintain and repair products and ensure they reach customers in good working order.

Chapter 9 provides an overview of the socio-economic and environmental impact of pico-solar products, explaining the health and safety benefits, how light can improve education, fight poverty and how access to electricity can contribute to local and national economies.

Chapter 10 identifies the challenges the sector faces in developing the largest potential market for pico-solar systems across rural Africa and Asia and discusses solutions, making recommendations for those seeking to facilitate market development.

Chapter 11 introduces case studies of innovative companies across Asia and Africa working to increase access to pico-solar systems.

Chapter 12 provides suggestions on further reading and an overview of the industry bodies, initiatives, programmes and companies operating in the pico-solar sector.

Acknowledgments

I'd like to thank:

Frank Jackson for his support, comments and input throughout the writing of this book.

Mark Hankins for permitting re-use of materials from his book, *Stand-Alone Solar Electric Systems,* particularly for Chapters 2 and 3.

Kat Harrison for writing Chapter 9 – The impact of pico-solar in the developing world.

Special thanks also to:

Marianne Kernohan, Peter Adelman, Daniel Davies, Zev Lowe for the Worldreader case study and all those who provided information for this book.

I'd also like to take this opportunity to thank Graham Knight for introducing me to the world of 'do it yourself' solar, Leo Blythe and the Kibera Community Youth Programme for working with me in Kenya in the early years and all the staff at SolarAid and SunnyMoney (past and present).

Last, but not least, thank you to my family, especially my wife Courtney (who is a great proof-reader) and my daughter, Molly, who was born while I was writing the early chapters.

List of Abbreviations

AC	alternating current
AGM	absorbed glass matt
Ah	amp-hour
BHAG	big hairy audacious goal
BoP	base of the pyramid (also known as bottom of the pyramid)
c-Si	crystalline silicon
DC	direct current
DoD	depth of discharge
Isc	short-circuit current
LED	light-emitting diode
Li-ion	lithium ion
$LiCoO_2$	lithium cobalt oxide
LFP/$LiFePO_4$	lithium iron phosphate
mAh	milliamp-hour
NiCd	nickel-cadmium
NiMH	nickel-metal hydride
OLED	organic light-emitting diode
OPV	organic photovoltaic
PAYG	pay as you go
PSH	peak sun hours
PV	photovoltaic
SHS	solar home system
SLA	sealed lead-acid`
SoC	state of charge
STC	standard test conditions
Voc	open circuit voltage
Wh	watt-hours
Wp	watt-peak

1

Introducing Pico-Solar

This chapter defines the term pico-solar, introduces the main components which make up a pico-solar system and provides examples of the many different types of pico-solar systems which exist. It goes on to identify those parts of the world where everyday access to electricity is limited and outlines how the small amounts of electricity generated by pico-solar systems can have a big social impact and transform peoples' lives. The chapter concludes with an overview of the rapidly expanding pico-solar market.

Pico-Solar – A New Category of Solar Electric Power

The rapidly expanding pico-solar industry is using the power of the sun to bring small, but incredibly useful and often life transforming, amounts of electricity to millions of people across the world. The terms pico-solar, picoPV or micro-solar are often used interchangeably to define and categorise small solar electric products and systems that are generally understood to be powered by solar modules with a power output ranging from as little as 0.1 watt-peak (Wp) up to 10–15 Wp. These levels of power are significantly lower than off-grid solar home systems (SHS), which are often 30–50 Wp.

Pico-solar systems come in a wide range of different forms and sizes, from solar lanterns and charging systems (with integrated or separate solar modules) to power an increasing array of energy efficient appliances, such as mobile phones, radios, digital cameras and e-readers, to integrated systems which power appliances such as parking meters and electric fences. Pico-solar systems generate small, relatively safe, amounts of electricity which means they do not normally need to be installed by a trained solar technician. Customers do, of course, need to understand how systems operate and how to look after them. This book also provides examples of slightly larger systems which operate according to the same principles as pico-solar systems, but use solar modules larger than 15 Wp.

Pico-solar systems are increasingly used across the world from rural households located beyond the electricity grid in need of light, to people in need of power on the go to keep their tablet charged, to the parking meters at the heart of the world's cities.

While the amount of electricity generated by pico-solar systems is low, the rise of low energy lighting and portable appliances such as mobile phones mean that this small amount of power can be incredibly useful for people travelling or living without regular access to electricity. The impact, especially for the 1.6 billion people who are not connected to the electricity grid or enjoy only intermittent access, can be transformational. Recognising the importance of pico-solar to human life, the BBC, in collaboration with the British Museum, chose the pico-solar powered lamp as its '100th Object' in its series *The History of the World in 100 Objects*. A UN report on how to eradicate poverty and transform economies for a post-2015 development agenda, meanwhile, has confirmed that pico-solar lights can save lives, reduce expenses and foster growth.

Pico-solar products have many advantages over more traditional solar systems. For example, they are often 'plug and play' – they do not require a solar technician to be installed and are relatively maintenance free. Crucially, however, pico-solar products are generally far less expensive than larger solar systems, making them more affordable and accessible for many across the world. Pico-solar systems, sold on the market for household use, typically range in price from around USD 10 for an entry level study light up to USD 150 for larger, multi-functional, systems. As technology improves and becomes more efficient, prices are also continuing to fall, while performance and product lifespans improve.

The past five years have seen a dramatic rise in the number of pico-solar products available on the market. In particular, there has been a rise in the field of off-grid power and lighting, which includes solar lanterns designed to offer a clean, safe alternative to kerosene lights. Figures 1.1 a, b and c show examples of some of the different types of pico-solar lights and systems available today. Figure 1.2 provides an indication of how the market has grown for pico-solar lighting devices in Africa, which is home to over 110 million un-electrified households.

This book is for anyone interested in learning about the growing pico-solar sector, from practitioners, manufacturers and retailers to policy makers, students, customers and socially driven eco-warriors. It provides the reader with a comprehensive overview of the pico-solar sector in twelve chapters which:

- explain what solar energy is and how it is used to generate electricity;
- cover each component which make up a typical pico-solar system – the solar module, the battery, circuitry and the appliances which the systems can power;
- discuss quality assurance issues and international standards;
- explain how to test products for quality and performance;
- provide guidance on how to use and maintain systems so as to maximise performance and lifespan;
- describe and provide evidence of the social impact systems, in particular lighting, are having on un-electrified households;
- provide an overview of the key challenges facing the market as well as the solutions, with a particular focus on how to reach and serve the large populations living on low incomes in communities with limited infrastructure;
- introduce a number of case studies from around the world where companies are bringing pico-solar systems to the market;
- provide the reader with resources and links to find out more about this growing sector.

Figure 1.1 Selection of pico-solar lights and charging systems, principally designed for use in areas without access to electricity, such as rural Asia and Africa. The product on the left has a separate 5 Wp solar module and can run multiple lights, charge phones and play radios. The product on the right has a separate 2.5 Wp solar module to run a single light and charge small appliances, such as mobile phones. The smallest product, in the centre, is an example of an entry level study light with an integrated solar module. Entry level study lights typically emit 20–30 lumens. Some larger, more powerful, pico-solar lights emit over 300 lumens. The World Bank/IFC Lighting Global initiative estimates that the number of manufacturers making pico-solar lights for markets across Africa and Asia has increased from just 20 in 2008 to over 80 in 2012.

Source: © David Battley

Pico-Solar Components

While pico-solar systems come in a range of shapes and sizes, a typical system is made up of the following components:

- **Solar module** (uses the sun's light to generate electricity);
- **Rechargeable battery** (stores electricity for use when needed);
- **Charge control circuitry** (protects the system from overcharging and deep discharge);
- **Power outlets** (connects appliances such as radios or phones);
- **Lighting** (incorporated as a key function in many systems).

The main technologies used in pico-solar system components are summarised in Table 1.1 below, and discussed in detail in Chapters 3–5.

Figure 1.1a Pico-solar systems are used to provide power for a wide range of applications. Figure 1.1a is a solar module integrated into a parking meter, which is becoming an increasingly common sight in cities across the world. The city of Portland, Oregon, for example, has installed 1,363 parking meters with an integrated 10 Wp solar module and sealed lead-acid battery, which needs replacing every 5–7 years. Solar parking meters have reduced the amount of waste produced by the city, which previously relied on disposable batteries that needed replacing each year.

Source: John Keane

Figure 1.1b A small portable charger with an integrated battery, PV module and LED light, also recharges mobile phones. Research by GSMA estimates there are around 600 million mobile users across the world without access to electricity, which means many have to travel long distances to the closest shops with electricity and have to pay to charge their phones. An estimated USD 10 billion a year is spent on charging phones in this way. Owning a pico-solar phone charger can therefore save people a lot of time and money.

Source: waka waka

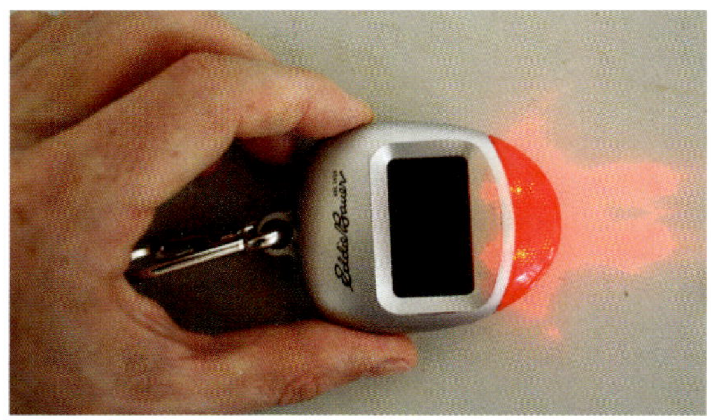

Figure 1.1c A small, pico-solar powered, bike light. This LED rear light has a small, integrated, thin film solar module (approximately 0.1 Wp) and an internal 2.4V rechargeable battery.

Source: John Keane

Table 1.1 Pico-solar system technologies

Solar PV Modules	Batteries	Lamps
Thin film (various)	Lithium ion – principally lithium iron phosphate ($LiFePO_4$)	Light-emitting diode (LED)
Polycrystalline	Nickel-metal hydride (NiMH)	Compact fluorescent lamp (CFL)
Monocrystalline	Sealed lead-acid	

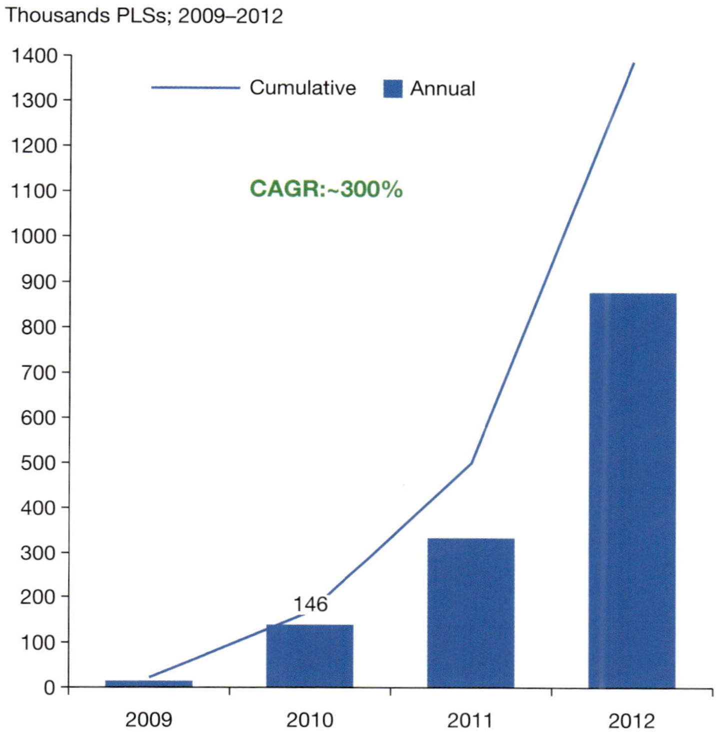

Figure 1.2 The sale of pico-solar lights in Africa, approved for quality by the Lighting Global programme, has increased significantly between 2009 and 2012. This trend is set to continue.

Source: Lighting Africa (2013)

The Multibillion Dollar Pico-Solar Market

While there is a growing market for portable pico-solar chargers which can charge up appliances such as phones for people travelling with limited access to electricity, the largest potential market for pico-solar products is the 1.6 billion people on the planet who live without access or have limited access to the electricity grid. The UN estimates that USD 23 billion is spent annually on kerosene for lighting, with others estimating that a further USD 10 billion is spent by people who have to pay to recharge their phones. Pico-solar chargers can therefore save people a great deal of money. Figure 1.4 shows that the majority of the world's un-electrified populations live in Asia and Africa. Figure 1.5 is a composite picture of the world at night which shows these parts of the world in darkness.

Most of the dark regions on the night map are located in the less developed regions or 'emerging economies' of Africa, Asia and South America, where the electricity grid often serves only a small proportion of rural populations. For many who do have access, the reliability of the grid is often poor, with frequent blackouts. Electricity can also be prohibitively expensive for many, forcing people to rely on kerosene and candles for lighting.

This situation is not set to change anytime soon. In Africa, a Lighting Africa report highlights that even the most optimistic projections show that the electricity grid will not expand quickly enough to keep up with population growth such that the approximately 110 million households without electricity today may

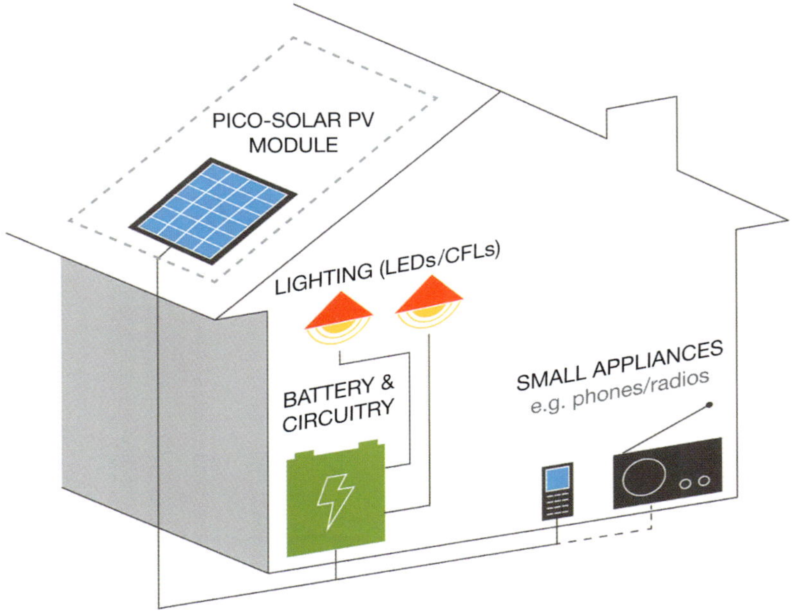

Figure 1.3a Pico-solar systems are smaller than traditional SHS (dotted line on roof indicates larger module size of SHS – also see Figure 1.3b). They can be used to power an increasing range of low power consumption appliances and lights. Pico-solar systems operate according to the same principles as SHS, but typically incorporate charge control circuitry into the main system housing as opposed to a separate charge controller unit. Pico-solar systems come in a wide range of different shapes and sizes, many of which are designed to be portable.

Source: John Keane

Figure 1.3b Large solar module dwarfing a pico-solar light and module.

Source: © Charlie Miller

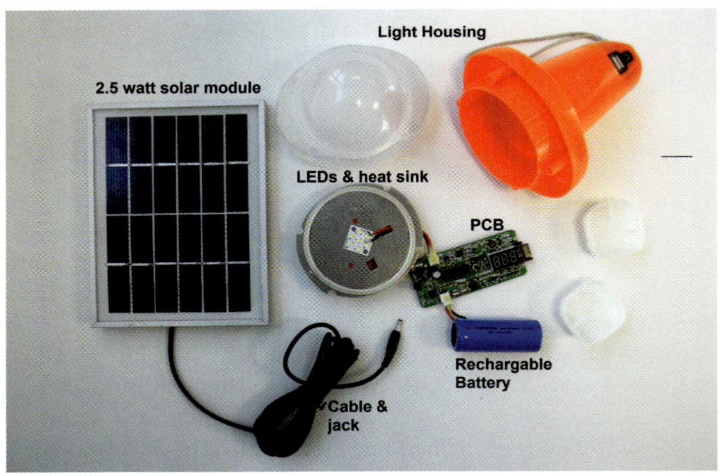

Figure 1.3c A solar light and phone charger taken apart to show the main components. The Printed Circuit Board (PCB) includes charge control circuitry designed to protect and manage the battery.

Source: John Keane

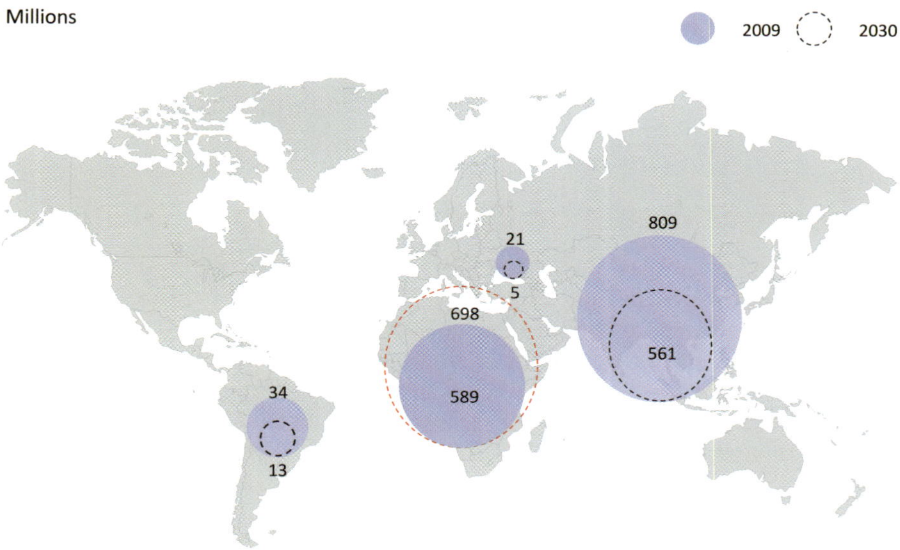

Figure 1.4 Map demonstrating that the largest un-electrified populations of the world in 2009 were in Asia (809 million people), Africa (589 million people) and South America (34 million people). Estimates for 2030 show that while the numbers are set to fall in Asia, populations without access to electricity are set to rise in Africa, which will then represent the largest un-electrified population in the world. Populations without access to electricity typically pay more for basic energy services, which are of lower quality, than those living on the grid.

Source: Electricity Access Database (International Energy Agency); Dalberg analysis Lighting Africa Market Trends Report 2012

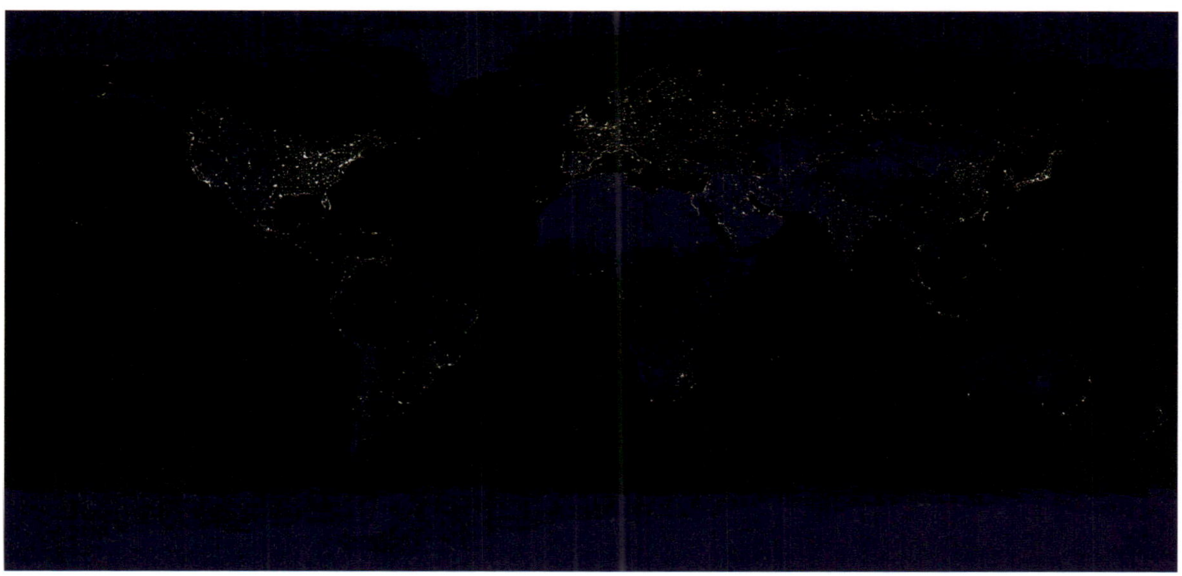

Figure 1.5 A composite image of the world at night which shows much of Africa, South America and Asia where access to electricity and electric lighting is limited, in darkness.

Source: http://eoimages.gsfc.nasa.gov/ve//1438/land_lights_16384.tif
Data courtesy Marc Imhoff of NASA GSFC and Christopher Elvidge of NOAA NGDC
Image by Craig Mayhew and Robert Simmon, NASA GSFC

increase to 150 million by 2030. The good news is that many of the regions of the world that lack access to electricity, benefit from high levels of solar radiation throughout the year (see Figure 1.6). This means that these sun-rich, but electricity poor, parts of the world are often extremely well located and suited to make effective use of solar energy to help meet electricity needs.

Kerosene's Impact on Global Warming

Environmental scientists have recently calculated that the unburnt black particulate from kerosene lighting contributes exponentially to global warming. Indeed, kerosene lamps account for as much as 3 per cent of global black carbon emissions. The black carbon effect is different to that of ordinary CO_2. Black carbon (soot) from kerosene lamps hangs in the air, where it reflects the sun and causes atmospheric temperature increases that directly contribute to global warming. The scale of the problem is much greater than previously realised. During its short atmospheric lifetime (a matter of days), 'one kg of black carbon produces as much "positive forcing" (the measure for atmospheric warming) as 700 kg of carbon dioxide (CO_2) does during 100 years' (Lam *et al.*, 2012: 4; Bond *et al.*, 2011). Black carbon is also said to contribute to global dimming, where soot and other particles absorb solar radiation and prevent it from reaching the earth's surface, causing cooling and potentially masking the effect of greenhouse gases on global warming.

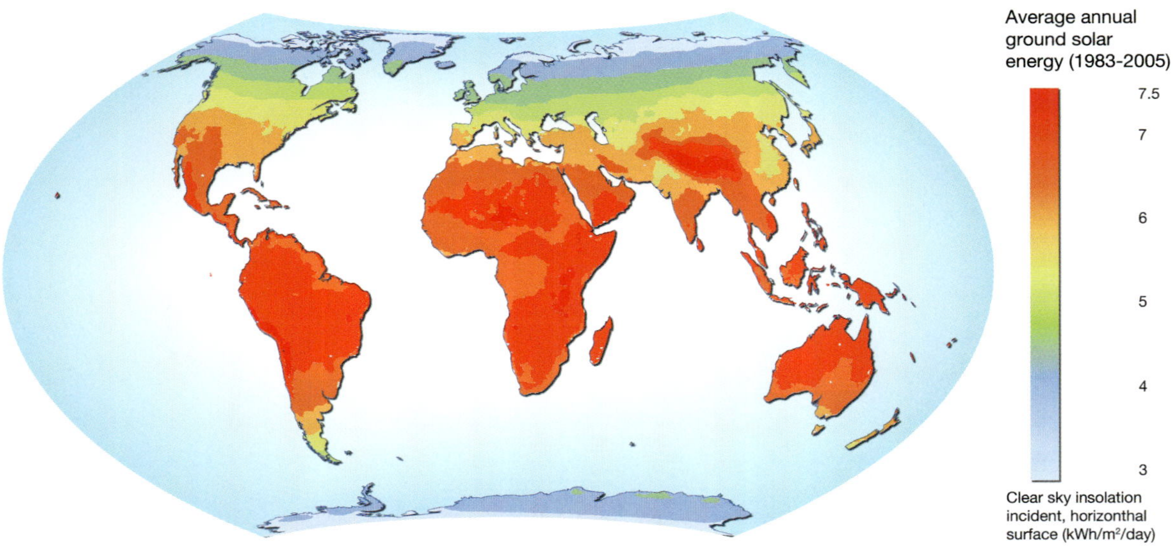

Figure 1.6 Incident solar radiation map, based on meteorological data collected by NASA, shows how much solar energy is received across the world. Many parts of the world with poor access to electricity across Asia, Africa and South America are sun-rich and well positioned to use solar power as an energy solution. Radiation levels and day lengths in many of these areas positioned relatively close to the equator do not vary significantly between summer and winter months.

Source: http://www.grida.no/graphicslib/detail/natural-resource-solar-power-potential_b1d5 Designer: Hugo Ahlenius, UNEP/GRID-Arendal. NASA. 2008

Figure 1.7 Kerosene lamps are a dangerous fire hazard, and are polluting, costly and emit low levels of light. Recent research by the US department of energy demonstrates that fuel-based lighting is up to 150 times more expensive than efficient electric lighting when the energy input–light output ratios are considered. Yet this type of crude kerosene lamp is still being used as the only source of light by tens of millions of the poorest people across the world. Kerosene lamps produce large quantities of smoke which are harmful to health – note the fumes in this photo. A Lighting Africa report cites a study which estimates that people who breathe kerosene fumes inhale the toxic equivalent of the smoke from two packets of cigarettes a day. In short, this type of light is better suited to centuries gone by, not the twenty-first century.

Source: John Keane

Figure 1.8 (below) The majority of households in rural Africa are not connected to the electricity grid. It is extremely common to see households located on main roads next to electricity lines but which remain unconnected. Note the electricity cables at the top of this picture which pass right by the houses, leaving the inhabitants without electricity. There are many reasons why small households do not connect to the electricity grid. The main ones are that the cost of connecting is too high and that people are not able to afford frequent electricity bills. In many parts of the world, households which can afford to connect to the grid are often subjected to frequent and lengthy power cuts.

Source: John Keane

Figure 1.9 Children light candles in a school in Malawi during a power cut. This image shows how difficult it is to study by candlelight. Power cuts are a major problem in many parts of the world. Many schools which do not enjoy access to electricity are forced to rely on kerosene lights or battery powered torches to facilitate evening study. This is expensive and the lighting is often of poor quality – which means many students are unable to study during the evening – a particular problem across equatorial Africa and Asia, where days and nights are often of equal length.

Source: SolarAid

Figure 1.10 Two students studying with one pico-solar light. The light is bright, constant, safe and can be recharged each day for free using sunlight. Just being able to study in the evening improves the chances of students across the world to gain a better education – something many of us take for granted.

Source: SolarAid

Power to Transform Lives

As pico-solar products are often portable and easy to use, like the mobile phone, they can be distributed quickly and can be put to good use immediately. As an example, a rural African household today, which is reliant on a candle or crude kerosene lamp for evening light, as shown in Figure 1.7, can now purchase some pico-solar lighting products for less than USD 10 to enjoy access to electric lighting and small, but useful, amounts of electric power.

Another small, portable electric device, the mobile phone, has arguably had a greater developmental impact on the world in the last 15 years than any single aid intervention. Expanding the mobile phone network is relatively inexpensive and easy to do, such that it has expanded far more rapidly than the electricity grid.

Figure 1.11 Children playing with disposable batteries which have reached the end of their life in Rema, Ethiopia. In areas without access to electricity, many people across the world resort to the use of disposable batteries to power torches and radios. It is expensive to continually purchase batteries, many of which last only for a matter of days before they need to be disposed of. The challenge is that many parts of the world do not have the infrastructure to dispose of batteries, which means batteries are often left to degrade and pollute the ground and expose children to dangerous chemicals. In parts of northern Argentina, villagers attempt to protect the ground from the dangers of battery pollution by incorporating batteries into homemade bricks and building materials.

Source: © Samson Tsegaye

Figure 1.11a Battery recycling box in Rome, Italy. Unfortunately, this is not a common sight in areas across the world without electricity access.

Source: John Keane

Pico-solar systems have the potential to have a similar transformative impact on quality of life as the mobile phone. They are also similar to mobile phones in that they do not require a physically connected network in order to operate. All they need is the sunshine, and with that they can generate the electricity needed to recharge phones and light homes. Bill Gates, co-founder of Microsoft and the Gates Foundation, has been quoted as saying, 'If you could pick just one thing to lower the price of, to reduce poverty, by far you would pick energy'. Pico-solar systems can help reduce household expenditure on kerosene, candles, batteries and phone charging. The development sector is taking note. The positive impact pico-solar can have on health, education, household safety, disposable income and the environment is discussed in detail in Chapter 9.

Overcoming Traditional Barriers to Solar Uptake

While there is a clear argument in favour of the use of solar power in the sun-rich areas of the world, the uptake of traditional SHS across much of the world has not been as rapid as was perhaps both hoped and expected. Damian Miller, in his excellent book, *Selling Solar*, states that:

> despite the fact that solar photovoltaic technology was proactively introduced to many emerging markets in the 1980s, its uptake was disappointingly slow. By the turn of the century only an estimated 1 million households were using a solar system for electricity, accounting for no more than 0.25% of un-electrified households globally. (p. 3)

While the uptake of solar systems is increasing, the two key barriers identified which hamper the rapid expansion of traditional solar power in emerging markets remain relevant:

- the absence of consumer finance to make solar more affordable; and
- the absence of a market infrastructure to make solar more widely available.

While traditional solar PV systems can save consumers a great deal of money over years of use, the upfront cost involved in purchasing a system is often a barrier that prevents many people from adopting solar power. Furthermore, with relatively few people adopting solar in emerging markets, only a small number of businesses are trying to supply it. Small, less expensive, pico-solar systems have the potential to address both of these barriers:

- Lower costs mean there is less need for consumer finance, making solar more affordable and accessible for many, overcoming some of the financial barriers.
- It is relatively easy for non-specialist shops and distributors to physically stock and retail pico-solar products which are often portable and do not need technicians to install; many leading entry level pico-solar lights are so small that it is possible to fit tens of thousands in one shipping container.

The pico-solar sector does face many challenges of its own, however, which are discussed in detail in Chapter 10.

Trends

The pico-solar industry is a fast moving sector which has seen explosive growth in recent years. This growth is set to continue, with prices of key components falling and performance of new technologies, in particular light-emitting diodes (LEDs) and lithium iron phosphate (LFP) batteries, improving. Ongoing innovations in technology, business models and the ever increasing array of energy efficient appliances that can be powered through pico-solar are further cause to be optimistic about future growth.

The key components that make up a pico-solar system, namely solar PV modules, LED lighting and new battery technologies, have seen consistent and often dramatic price reductions over time. While the future price of PV modules remains uncertain, the indications are that price reductions for LED lighting and LFP batteries are set to continue for the foreseeable future. There is, therefore, every reason to believe that the price of pico-solar systems will continue to fall as components drop in price and the industry achieves economies of scale as it grows. Component price reductions also enable manufacturers to increase the quality and performance of products without increasing the price to the customer.

It is worth noting that while pico-solar systems are less expensive than traditional systems, even prices as low as USD 10 for a study light are a barrier for many people living in poverty, and price reductions are always welcome.

While the price of pico-solar systems is set to fall, the price of kerosene has historically seen steady increases over time. It is therefore possible to say with confidence that pico-solar systems will become increasingly attractive options for households which traditionally rely on fuels such as kerosene for lighting.

Alongside continued reductions in price, the overall quality and performance of pico-solar systems is set to improve as LEDs and lithium-based batteries in particular continue to evolve. This means that, over time, the industry will be able to offer better products to customers at better prices. Indeed, this is already happening, with several leading manufacturers launching new generation product lines in 2012 with brighter LEDs, longer lasting batteries and longer warranties than previously offered, without increasing any of their prices. Better products at better prices can only mean more customers will be willing and able to purchase pico-solar products and benefit from access to electricity.

Maturing Market

The number of companies manufacturing, distributing and retailing pico-solar products across the world is increasing each year with more pico-solar products reaching off-grid households every day. Notwithstanding the current growth, it is estimated that less than 0.5 per cent of households in Africa own a pico-solar product while penetration rate of pico-solar lanterns and SHS in India, the country with the highest off-grid population in the world, is estimated at 4.5 per cent of households.

As the pico-solar market continues to develop, it will need to establish more developed distribution chains, finance solutions and after-sales service and brands which consumers recognise and trust. This is particularly challenging in hard-to-reach and service markets in parts of rural Africa, Asia and South America. These challenges are discussed in Chapter 11.

The United Nations General Assembly declared 2014–2024 as the Decade of Sustainable Energy for All. This declaration underlines the importance of energy issues for sustainable development and pico-solar solutions will play a central role in helping to achieve these goals. Chapter 12 provides an overview of the increasing number of industry bodies and initiatives which aim to support the development of the pico-solar sector and increase access to modern energy services.

2

The Solar Resource

This chapter introduces the reader to the basic principles of solar energy and how it can be harnessed to generate electricity. On completing this chapter, you will have a good understanding of what solar power is and how it works. The chapter also includes a short overview of what electricity is and how it is measured.

Solar Power – Putting the Sun's Energy to Work

The sun is a wonderful source of energy and is vital to life on our planet. Without it, we just would not exist. The amount of energy produced by the sun is immense: 3.8×10^{23} kW of power. While not all of this energy reaches the earth, we still receive a whopping 1.73×10^{16} kW, which is thousands of times more than enough to provide all of humanity's annual energy needs.Thomas Edison is often quoted as saying 'I'd put my money on the sun and solar energy. What a source of power! I hope we don't have to wait until oil and coal run out before we tackle that.'

The challenge is finding ways in which we can practically use this energy. Plants do this every day, converting sunlight into chemical energy through photosynthesis. The sun's heat is also put to use every day, not least to dry peoples' clothes and heat water tanks with solar thermal collectors. Industrial solar thermal systems have also been developed which harness the sun's heat to create steam to power electric turbines.

The focus of this book, however, is on solar photovoltaic modules (solar PV) which use the sun's light to generate electricity. Figure 2.1 below shows the transfer of energy through a pico-solar system, starting with the sun's rays, which solar PV modules convert into electric current which then flows into a rechargeable battery, charging it up. Here it is stored as chemical energy and available for use when needed to power lights and other appliances or 'loads'. Box 2.1 provides a short overview of what electricity is and the key terms which are used when explaining it.

In order to get the most out of a pico-solar system, it is useful to understand more about the solar resource itself. This sections introduces the key principles of solar energy and explains key terms such as solar radiation; irradiance, solar incident angle and insolation.

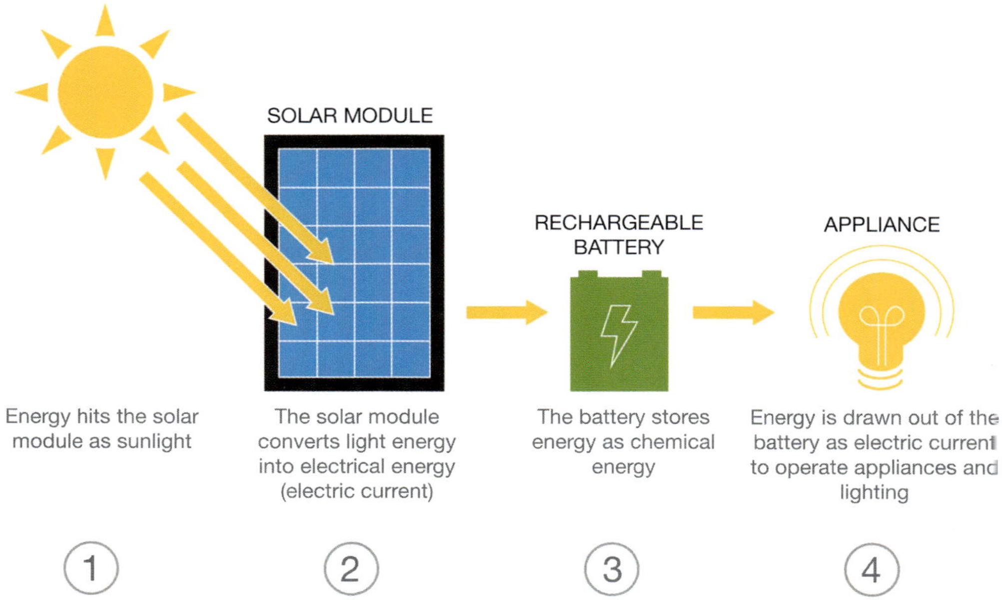

Figure 2.1 Energy is transferred from the sun as light energy and then converted into electrical energy which charges up a rechargable battery, where it is stored as chemical energy. The energy is drawn from the battery as electricity which is used to run electrical appliances.

Source: John Keane

Solar Radiation Principles

Sunshine reaches the earth as a type of energy called radiation. Radiation is composed of millions of high energy particles called photons. Each unit of solar radiation, or photon, carries a fixed amount of energy – see Box 2.1. Depending on the amount of energy that it carries, solar radiation falls into different categories including infrared (i.e. heat), visible (radiation that we can see) and ultraviolet (very high energy radiation). The solar spectrum describes all of these groups of radiation energy that are constantly arriving from the sun, and categorises them according to their wavelength. Different solar cells and solar energy-collecting devices make use of different ranges of the solar spectrum.

Solar energy arrives at the edge of the earth's atmosphere at a constant rate of about 1350 W/m^2 (watts per square metre): this is called the 'solar constant'. Due to absorption by the earth's atmosphere, the amount of energy that actually reaches the sun's surface is reduced to a maximum of about 1000 W/m^2. This means that when the sun is directly overhead on a sunny day, solar radiation is arriving at the rate of about 1000 W of power per square metre of the earth's surface (1000 W/m^2).

The number of daylight hours in countries located between and around the tropics remains fairly constant throughout the year, whereas day lengths vary considerably in countries in other parts of the world, such as Europe in the northern hemisphere, where winter days are quite short. This means that annual solar radiation levels are generally higher around the tropics than elsewhere in the world.

Box 2.1 Basic Energy and Power Concepts

Energy

Energy is referred to as the ability to do work. Energy can be measured in watt-hours (Wh) which is a useful way of measuring electrical energy. One watt-hour is equal to a constant 1 watt supply of power supplied over 1 hour. If a light is rated at 2 watts, in 1 hour it will use 2 Wh. In four hours it will use 8Wh of energy.

Power

Power is the rate at which energy is supplied (or energy per unit time). Power is measured in watts (W).

Power conversions

watts × 746 = horsepower
watts × 1000 = kilowatt (pico-solar systems do not operate at this level of power, however).

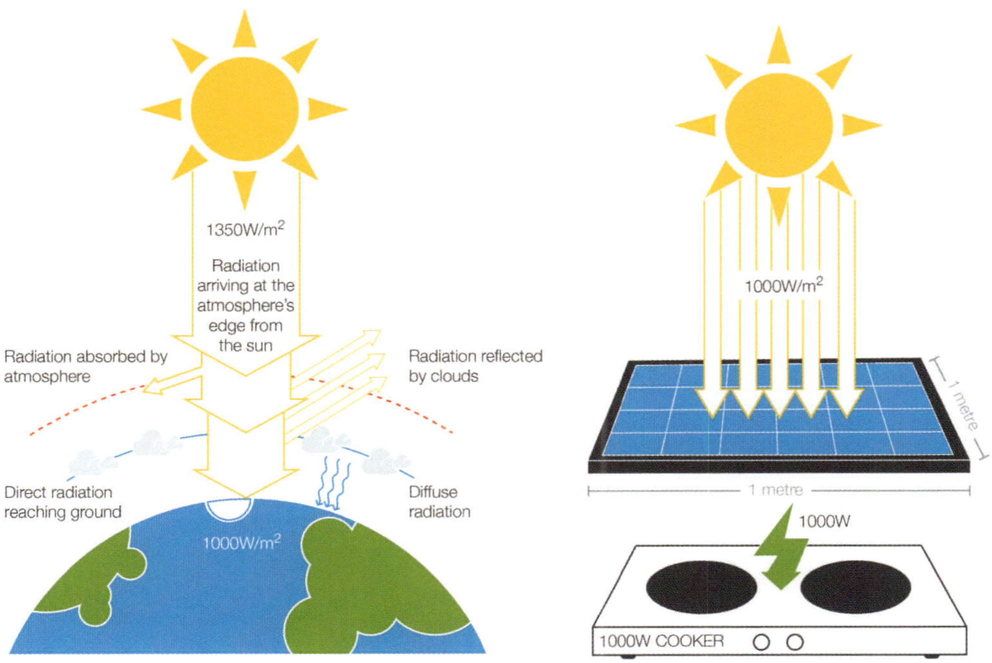

Figure 2.2 1350 W/m^2 of solar radiation arrives in the earth's atmosphere. Some of this radiation is absorbed by the atmosphere and reflected by clouds such that 1000 W/m^2 reaches the earth's surface (left). A reasonable PV module is capable of converting about 10 per cent of this energy. An indication of how much power 1000 W/m^2 actually is shown on the right by relating it to the rating of a small electric cooker.

Source: adapted from Hankins (2010)

Direct and Diffuse Radiation

Solar radiation can be divided into two types: direct and diffuse. PV modules use both and most of the time it does not matter very much for the types of systems discussed in this book whether the radiation is direct or diffuse as long as overall radiation (called global radiation) is high enough through the day. Direct radiation comes in a straight beam and can be focused with a lens or mirror. Diffuse radiation is radiation reflected by the atmosphere, or radiation scattered and reflected by clouds, smog or dust. Clouds and dust absorb and scatter radiation, reducing the amount that reaches the ground. On a sunny day, most radiation reaching the ground is direct, but on a cloudy day up to 100 per cent of the radiation is diffuse. Together, direct radiation and diffuse radiation are known as global radiation.

Radiation received on a surface in cloudy weather can be as little as one-tenth of that received in full sun. Therefore, solar systems must be designed to guarantee enough power in cloudy periods and months with lower solar radiation levels. At the same time, system users must economize energy use when it is cloudy.

Annual and even monthly solar radiation is predictable. Factors that affect the amount of solar radiation an area receives include the area's latitude, cloudy periods, humidity and atmospheric clarity. At high intensity solar regions near the equator, solar radiation is especially affected by cloudy periods. Long cloudy periods significantly reduce the amount of solar energy available. High humidity absorbs and hence reduces radiation. Atmospheric clarity, reduced by smoke, smog and dust, also effects incoming solar radiation. The total amount of solar energy that a location receives may vary from season to season, but is quite constant from year to year.

Figure 2.3 Direct and diffuse radiation. Cloud cover reduces the amount of direct radiation reaching the earth's surface.

Source: adapted from Hankins (2010)

Solar Irradiance

Solar irradiance refers to the solar radiation actually striking a surface, or the power received per unit area from the sun. This is measured in watts per square metre (W/m^2). If a solar module is facing the sun directly (i.e. if the module is perpendicular to the sun's rays) irradiance will be much higher than if the module is at a large angle to the sun.

Solar Incident Angle

The angle at which the solar beam strikes the surface is called the solar incident angle. The closer the solar incident angle is to 90°, the more energy is received on the surface. If a solar module is turned to face the sun throughout the day, its energy output increases.

Insolation

Insolation (a short way of saying incident solar radiation) is a measure of the solar energy received on a specified area over a specified period of time. Meteorological stations throughout the world keep records of monthly solar insolation.

Peak sun hours (PSH) is defined as the equivalent number of hours each day when solar irradiance averages 1000 W/m^2. For example, a site that receives 6 PSH a day, receives the same amount of energy that would have been received if the sun had been shining for 6 hours at 1000 W/m^2. In reality, irradiance changes throughout the day, with less irradiance in the early morning and evening

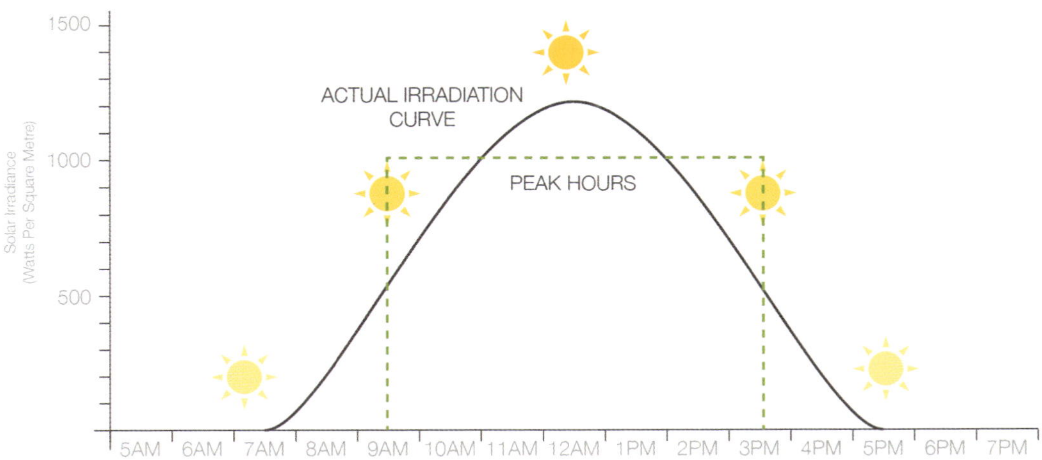

Figure 2.4 Solar irradiance in watts per square metre (W/m^2), received over time on a flat surface in an equatorial region. In the morning and late afternoon, less power is received because the flat surface is not at an optimum angle to the sun and because there is less energy in the solar beam. At noon, the amount of power received is highest. The actual amount of power received at a given time varies with passing clouds and the amount of dust in the atmosphere.

Source: adapted from Hankins (2010)

Figure 2.5 The solar incident angle. More energy is collected for conversion into electricity when the solar module faces the sun.

Source: adapted from Hankins (2010)

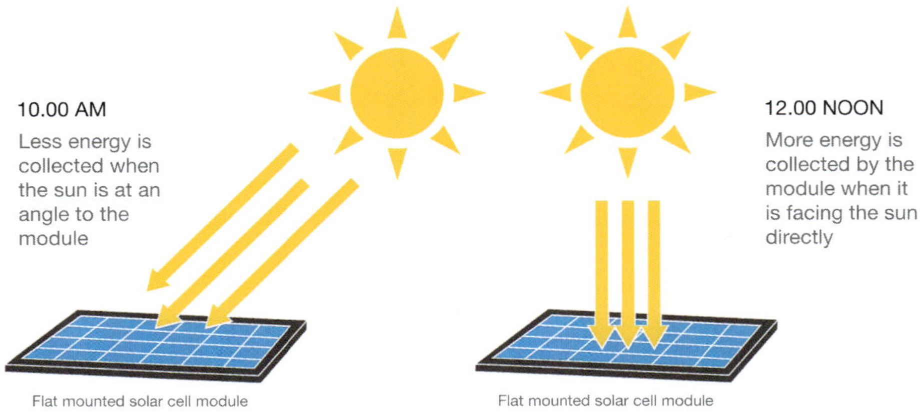

than in the middle of the day. One PSH can be achieved by irradiance of 500 W/m^2 for two hours, or 250 W/m^2 for four hours and so on. Figure 2.4 provides an example of a solar energy site with 5–6 PSH, with irradiance above 1000 W/m^2 for about 3 hours, above 500 W/m^2 for about 4 hours and above 200W/m^2 for about 2 hours. PSH per day can vary considerably from season to season. Taking Addis Ababa in Ethiopia as an example, this location may record over 6 PSH per day (6 kWh/m^2/day) in October and November, but may drop below 4 PSH in July and August.

Electricity Basics

Electricity can be defined as the movement of charged particles or electrons which flow as electric current through an electrical conductor, such as metal wires.

When electric current flows in a single direction, this is called direct current (DC). Direct current is produced by solar cells and batteries and can be used to run electric appliances. Alternating current (AC) refers to electric current which periodically changes direction and is used to distribute power across large distances more efficiently. It is possible to convert DC into AC by using an inverter and to convert AC into DC using a rectifier. This book only deals with systems and appliances which operate with DC.

Figure 2.6 Symbols used for direct current and alternating current.

Source: John Keane

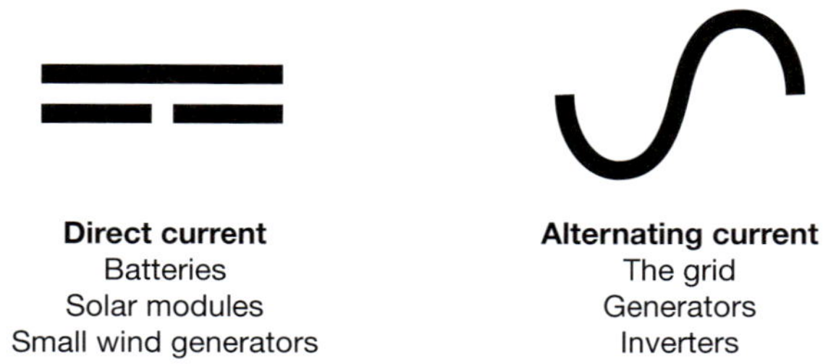

Box 2.2 Units of Measurement

Volts (V): Pressure difference between two points (electrical potential difference) which causes electric current to flow. (Symbol used: V).

Amps (A): The measurement for electric current, which is the flow of electrons or charged particles through an electrical conductor. (Symbol used: I).

Amp-hour (Ah): the amount of charge in a battery; one amp-hour is equivalent to one amp of current flowing for one hour. Used to indicate battery capacity and also to explain amount of charge put into or drawn from a battery, e.g. a solar module which provides 3A of current for 3 hours, provides 9Ah. As pico-solar systems use relatively small solar modules and batteries, it is common to see reference to mili-amp-hour (mAh) – 1Ah = 1000mAh).

Watts (W) A watt is a measurement of electrical power defined as the rate at which work is done when one ampere flows through an electrical potential difference of one volt.

Ohms (Ω) Unit of resistance equal to the resistance of a circuit in which a potential difference of one volt produces a current of one amp.

Watt-hour (Wh) Energy measure, calculated by multiplying power (watts) by hours.

Watt-peak (Wp) Maximum power output generated by a PV module under Standard Test Conditions (STC).

Basic calculations

Power (watts) = volts × amps (P = VI)
Watts × hours = Watt-hours (Wh)
Amps × hours = Amp-hours (Ah)

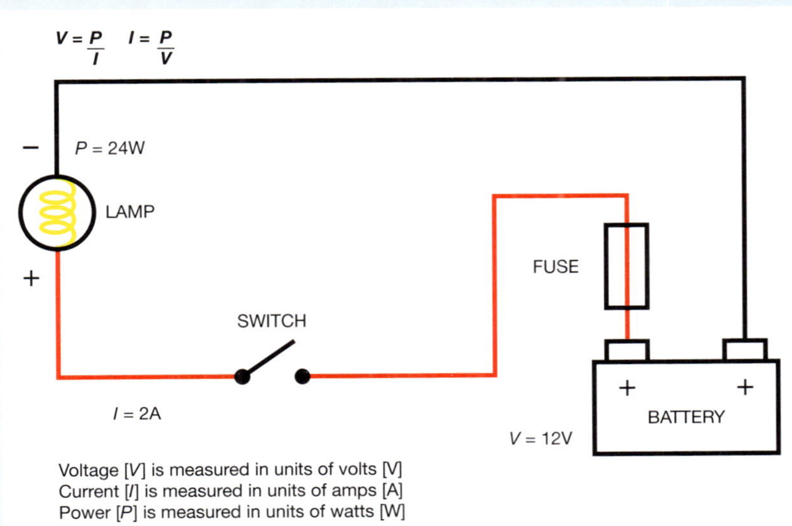

Figure 2.7 Voltage, current and power within the context of a simple circuit. This simple circuit has electricity flowing at a rate of 2A from the positive terminal of a 12V battery, through a fuse to power a lamp. Power is measured in watts. In this circuit the power involved is 24 watts, calculated by multiplying 12V by the 12A of current. The fuse is there to protect the lamp from any surges in electric current. The switch is used to create a break in the circuit and to stop the flow of electricity in order to turn the lamp off.

3

Solar PV Cells and Modules

Photovoltaics (PV) is the term used to describe the solar technologies that convert sunlight into electricity. It is common to use the term *solar photovoltaics* shortened to *solar PV* or just *PV*. This chapter explains how solar PV cells and modules work and introduces the key ratings which are used to measure the energy output of cells and modules. It also shows the reader how to carry out simple measurements with a multimeter. The so-called IV curve is then explained together with how energy output is affected by different conditions such as temperature and radiation. The chapter concludes with an overview of the main PV technologies used within the pico-solar sector, price trends and a look at innovations in PV technology.

The Solar Cell

The basic unit of solar electric production is the solar cell. Light striking solar cells creates a current powered by incoming light energy. They produce electricity when placed in sunlight. Most solar cells do not get used up or damaged while generating electric power. Their life is limited only by breakage or long-term exposure to the elements. If a high quality solar cell module is properly protected, it should last for more than 25 years. Figure 3.1 below shows a solar cell generating electricity.

How Solar Cells Work

Solar cells rely on the special electric properties of silicon (and other semi-conductor materials) that enable it to act as both an insulator and a conductor. Specially treated wafers of silicon 'sort' or 'push' electrons dislodged by solar energy across an electric field on the cell to produce an electric current.

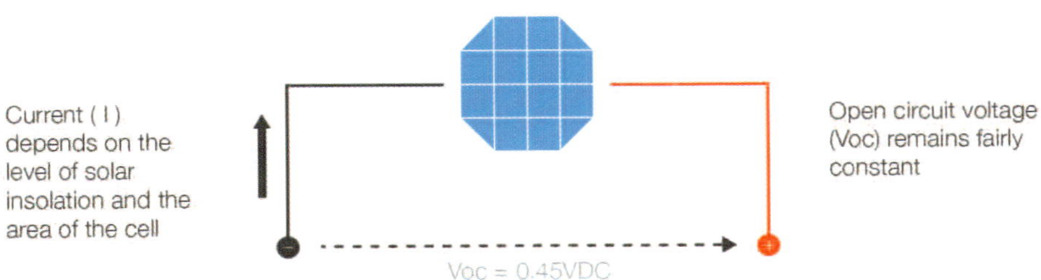

Figure 3.1 A solar cell producing electricity. In many pico-solar systems the modules consist of cells cut in half or in quarters. This reduces the surface area of the cell and thus the current output but not the voltage. The voltage output of half a cell is the same as the voltage output as a complete cell.

Source: John Keane

Solar radiation is made up of high energy sub-atomic particles called photons. Each photon carries a quantity of energy (according to its wavelength); some photons have more energy than others. When a photon of sufficient energy strikes a silicon atom in a solar cell, it 'knocks' the outermost silicon electron out of its orbit around the nucleus, freeing it to move across the cell's electric field, also called the p-n (positive-negative) junction. Once the electrons cross the field, they cannot move back. As many electrons cross the cell's field, the back of the cell develops a negative charge.

If a load is connected between the negative and positive sides of the cell, the electrons flow as a current. Thus, solar energy (in the form of photons) continuously dislodges silicon electrons from their orbitals and creates a voltage that 'pushes' electrons through wires as electric current. This process is illustrated in Figure 3.2. More intense sunlight gives a stronger current. If the light stops striking the cell, the current stops flowing immediately.

The Solar Module

The amount of current produced by a solar cell depends on its size and type. Silicon-type solar cells generate a potential difference (voltage) of between 0.4 and 0.5 V in normal operation.

Solar cells must be connected in series to increase voltage to a useful level – see Figure 3.3. For most tasks, it is not convenient to use single solar cells because their output does not match the load demand. For example, one cell cannot power a radio if the radio requires current at 3 V and the cell produces a voltage of only 0.5 V. Five cells in series are enough to power a calculator of 2 V and 18 cells are normally required to charge a 6 V battery. Thus solar cells are arranged in series to increase voltage, and the number of cells depends on the application. Moreover, crystalline solar cell wafers are fragile, so they must be protected from breakage and corrosion. For these reasons, solar cells are electrically connected in series, then packaged and framed in devices called photovoltaic modules.

The process of making solar cell modules from monocrystalline and polycrystalline silicon cells involves several steps. Once properly prepared and treated with anti-reflection coatings, solar cells are soldered together in series (i.e. the front of one cell is connected to the back of the next) and then mounted between

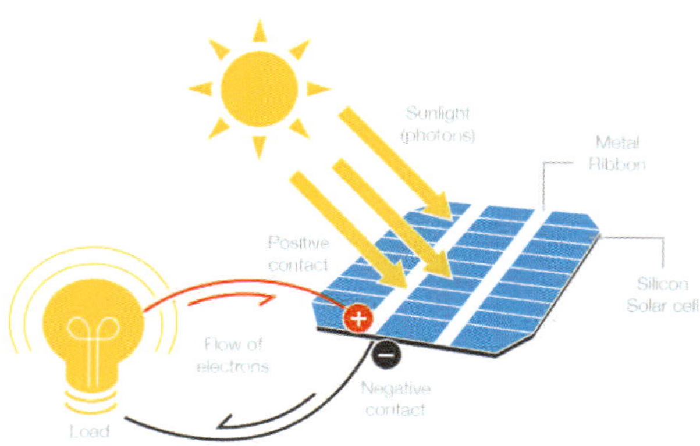

Figure 3.2 Sunlight hits the surface of the PV cell, which converts the light energy (photons) into electric current which flows to the terminals which are connected to a circuit. The electric current flows through the circuit and provides power to operate an appliance (load), which in this case is a lamp.

Source: adapted from Hankins (2010)

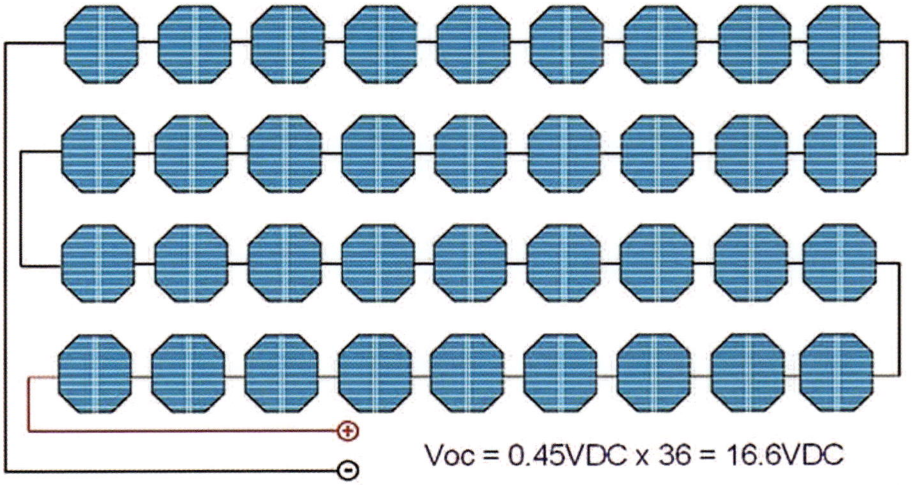

Figure 3.3 Thirty-six solar cells wired together in series so as to achieve a higher voltage. In this case the Voc is 16.6 V, which is high enough to charge up a 12 V battery.

Source: John Keane

Figure 3.3a A solar module made up of solar cells wired together.

Source: John Keane

glass and plastic. The process by which monocrystalline or polycrystalline solar cells are sealed between glass and plastic is called 'encapsulation'.

During encapsulation, cells are sealed at high temperature between layers of plastic (a special type called EVA plastic) and glass in such a manner that air or water cannot enter and corrode the electrical connections between the cells. Modules are then cased in metal or plastic frames to protect their edges and to protect them from twisting. The frame may have holes drilled in it for easy mounting and connection points for earthing/grounding cables.

Positive and negative electric contacts from the cells, either terminal screws or wires, are fixed on to the back of the module.

Module Ratings

All solar cell modules are rated according to their maximum output, or 'peak power'. Peak power, abbreviated Wp (watt-peak), is defined as the amount of power a solar cell module is expected to deliver under STC. The module's power rating in peak watts, the short-circuit current, open circuit voltage and maximum outputs should be specified prominently on the module's label. See Figure 3.4. These terms are explained in the next section. Modules almost always produce less power than their rated peak power in field conditions. The module should also display an acceptable international certification such as IEC 61215 (for crystalline modules) or IEC 61646 (for thin film modules).

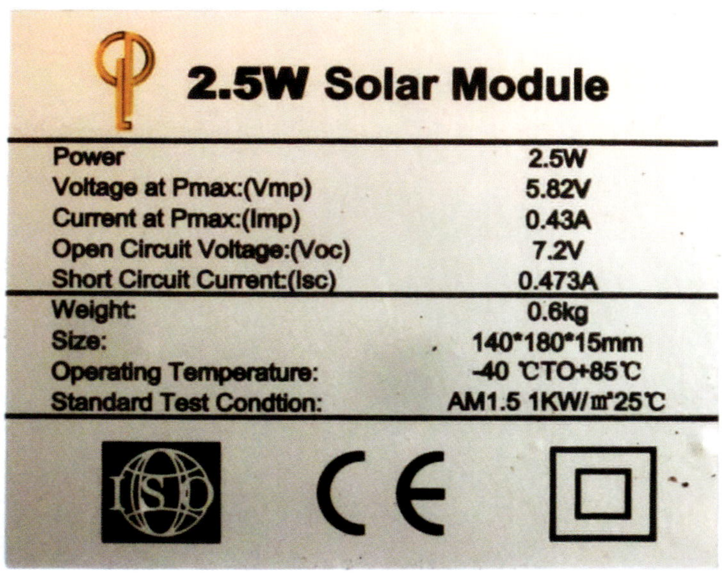

Figure 3.4 The back of a pico-solar module provides information such as power rating and operating specifications under test conditions. Always check what type of guarantee the module offers. Modules should come with at least a 5-year warranty. Many good crystalline modules have 25-year warranties.

Source: John Keane

Box 3.1 Standard Test Conditions and Measurements

STC allow manufacturers and consumers to compare how different PV devices perform under the same conditions. A number of laboratories and centres test modules and cells for manufacturers. STC are set at:

Solar Irradiance:	**1000 W/m^2**
Temperature:	**25°C**
Air mass:*	**1.5**

*(Air mass indicates how much radiation is absorbed by the atmosphere).
Note that modules and cells almost always produce less power under actual working conditions.
The following measurements are normally taken at STC and recorded on the back of the module:

- Voltage at Pmax (Vmp) is the voltage at which the module produces the greatest power.
- Open circuit voltage (Voc), is the voltage measured with an open circuit. It is measured with the solar cell in full sunlight using a voltmeter attached to the positive and negative leads of the module (see Figure 3.9).
- Current at Pmax (Imp) is the current at which module produces greatest power.
- Isc, the short-circuit current, is the current measured in full sunlight when the positive and negative wires are 'shorted'. This can be measured using a mulitimeter which is attached to the positive and negative leads of a solar module (see Figure 3.10).

Output of Solar Modules

The power output of a module depends on the number of cells in the module, the type of cells and the total surface area of the cells. Output varies according to changes in:

- solar radiation levels;
- angle of the module in relation to the sun;
- temperature;
- voltage at which the battery is drawing power from the module.

The I-V curve

I-V curves are used to compare solar cell modules and to predict their performance at various temperatures, voltage loads and levels of insolation. Each solar cell and module has its own particular set of operating characteristics. At a given voltage, a module will produce a certain current. These properties are described by the current-voltage curve, better known as the I-V curve. Figure 3.5 shows a typical I-V curve for a solar cell. The left-hand side (I) gives the current cell produces depending on voltage. The bottom side gives the voltage produced by the cell at various currents. At each point along the line it is possible to determine the power of the module by multiplying the current by the voltage.

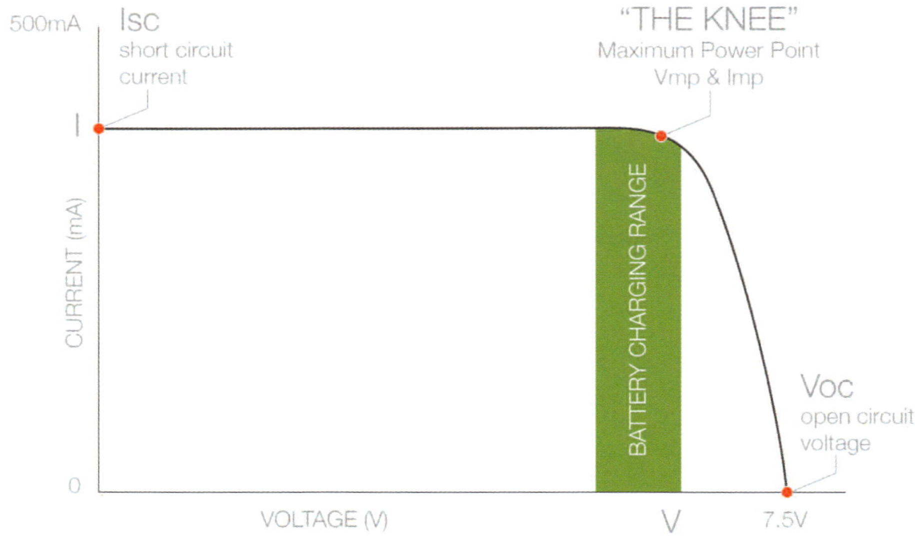

Figure 3.5 Silicon solar cell I-V curve which shows the maximum power point 'the knee' of the curve. Isc is the point where the curve crosses 0 volts. This is the maximum current that the cell or module is capable of producing. (See Figure 3.10). Voc is the point where the curve crosses 0 amps. This is the maximum voltage that the module can produce on a sunny day (See Figure 3.8). The maximum power point is always found at the place where the curve begins to bend steeply downward ('the knee'). Crystalline modules have I-V curves that are more 'square' than thin film modules.

Source: John Keane

Pico-solar systems do not generally include charge control circuitry which enables the module to always function at the maximum power point (the knee on the I-V curve), such that the operating voltage of modules is determined by the battery's voltage. This basically means that the module will not operate at the maximum power point throughout a charging period and will vary according to the changing voltage of the battery.

Effects of Radiation Intensity on Module Output

Solar module output is determined mainly by the intensity of the solar radiation on a module. Figure 3.6 shows that module output is directly proportional to the solar irradiance. Halving the intensity of solar radiation reduces the module output by half. Lower radiation also lowers the voltage at which current is produced.

Cloud cover reduces the power output of a module to a third or less of its sunny weather output. During cloudy weather, the voltage of a module is also reduced.

Effects of Heat on Module Output

Unlike solar thermal devices, most solar PV modules produce less power as they get hotter. As the temperature increases, power output of monocrystalline solar cells falls by 0.5 per cent per degree centigrade (this is shown by the I-V curve in Figure 3.5). Thus, a 5°C rise in temperature will cause a 2.5 per cent drop in power output. When mounted in the sun, solar cell modules are usually 20°C warmer than the thermometer temperature. Note the differences in the I-V curves at various temperatures. At 75°C the current output of the module is much lower than at 0°C.

This is important because the temperature on some tin rooftops common across much of the world can reach 75°C, reducing the output of the module. For this reason, it is recommended that modules are not mounted directly onto

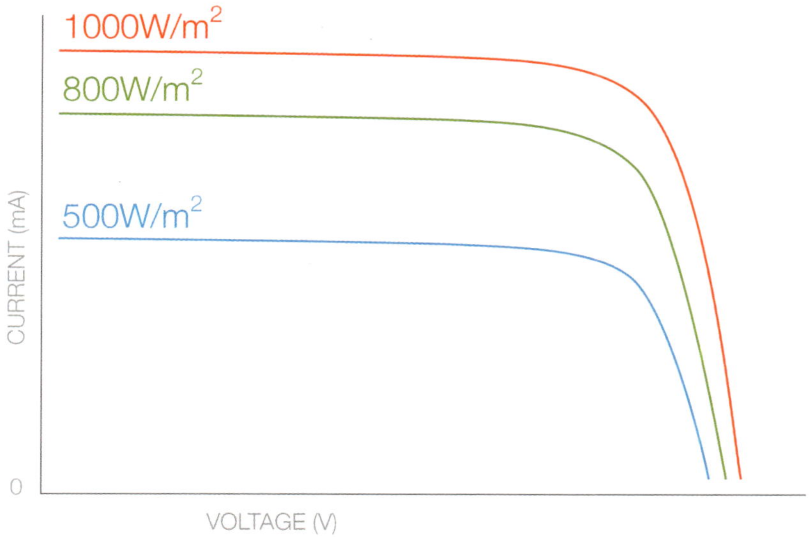

Figure 3.6 Effects of radiation intensity on module output

Source: adapted from *Lighting Global Technical Note*

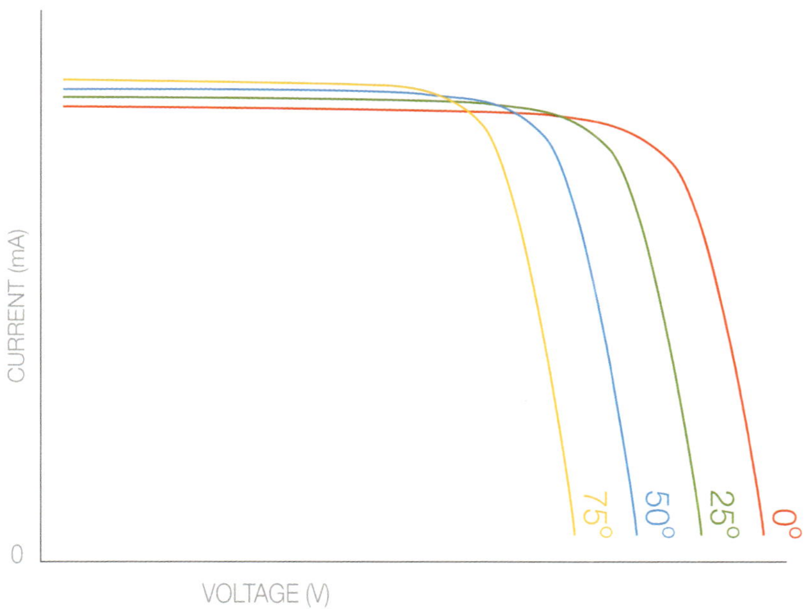

Figure 3.7 Effects of temperature on module output. Note: open-circuit voltage (Voc) reduces significantly as temperature increases.

Source: adapted from *Lighting Global Technical Note*

tin roofs and to position modules in places where they are cooled by airflow to keep output as high as possible

Effects of Shading on Module Output

Obviously, if a shadow falls across all or part of a module, its electric output will be reduced. In fact, even shading a single cell can considerably lower a crystalline module's output and possibly damage it. Damage occurs because the cells in a

Figures 3.8 and 3.9 In full sunlight, the module shows a Voc reading of 21.24 V (top). The Voc reading drops to 20.73 V when the module is in the shade (bottom).

Source: Frank Jackson

module are connected in series and they each must carry the same current. When one cell (or more) is shaded, it stops producing current and instead consumes current, converting it to heat.

Effect of Module Positioning on Output

If a single cell is shaded for a long time, it may cause the entire module to fail. Even a single tree branch, a weed or a bird's nest can shade one cell and cause electrical production to fall dramatically. Amorphous and multi-junction type modules are less affected by small shadows than crystalline-type modules. By

Figures 3.10 and 3.10a In full sunlight, the Isc reading is 0.305 A (top). This reading drops significantly to 0.071 A even when the module is only partly covered with shading (bottom). This illustrates the importance of keeping modules free of dirt, shade or anything that reduces the amount of sunlight reaching its surface.

Source: Frank Jackson

covering one cell of a crystalline module when measuring the current output the difference will be seen immediately.

Lighting Global has produced a detailed technical note which provides more information on testing module performance by measuring I-V curves. See www.lightingafrica.org

Types of PV Modules

Pico-solar PV modules are exactly the same as normal solar PV modules – they are just smaller and made up of fewer solar cells, which are often cut in half or quarters, to reflect the lower voltage and current levels required. There are three types of photovoltaic modules available for use in the pico-solar module market:

- Monocrystalline silicon;
- Polycrystalline silicon;
- Thin film: there are various different types of thin film. Amorphous silicon is the dominant thin film in the pico-solar. Cadmium telluride and copper indium gallium (di)selenide contain toxic hazardous materials and not generally found in pico-solar modules or recommended due to lack of infrastructure to dispose of safely.

General characteristics of different types of PV technology are shown in Table 3.1. The efficiencies used are approximations of those as measured in laboratory conditions. In reality, efficiencies are typically lower when measured in field conditions.

Monocrystalline Silicon

These modules are the most efficient on the market, approaching 15–18 per cent efficiency – when measured under STC in laboratories, but they are also the most expensive. Due to their efficiency, monocrystalline modules can be quite a bit smaller than other types of module.

Table 3.1 General characteristics of different types of PV technology

Technology	Lifespan (Years)	Efficiency	Disposal Issues	Notes
Monocrystalline Silicon	25 +	15–18%		Generally good quality, long-lasting, recommended
Polycrystalline Silicon	25 +	13–16%		Generally good quality, long-lasting, recommended
Thin Film				
Amorphous Silicon	20 +	6–12%		Output declines by up to 25% after first few months of use but on a good quality product the Wp value given on the label will be the output after this decline Lower efficiency means modules have larger surface area for same output as crystalline
Cadmium telluride			Require specialist recycling facilities	
Copper indium gallium (di)selenide			Require specialist recycling facilities	

The efficiencies used in this table refers to approximations of those as measured in laboratory conditions. In reality efficiencies are typically lower when measured in field conditions.

Figure 3.11 and Figure 3.11a Monocrystalline cells (top) and modules (bottom) can be recognised by their uniform dark blue/black background which is covered with a lighter grid. They can look somewhat similar to polycrystalline silicon modules. To be absolutely sure, refer to the relevant datasheets.

Source: JA Solar

Source: Frank Jackson

Polycrystalline Silicon

These modules are less efficient than their monocrystalline counterparts (13–16 per cent when measured under STC in laboratories) which means that a slightly larger module is needed in order to generate the same amount of power. They are, however, usually slightly less expensive.

Thin Film

Thin film modules are the least efficient type of PV module, ranging from around 6–12 per cent when measured under STC in laboratories, depending on which materials are used. They are, however, generally the cheapest.

Thin film modules can be made from a range of different materials, the main ones being:

- amorphous silicon (a-Si);
- cadmium telluride (CdTe);
- copper indium gallium selenide (CIS/CIGS).

Modules containing cadmium telluride and copper indium gallium selenide are not recommended in areas where proper disposal is not possible due to the toxicity of the active materials – in module form they are perfectly safe. Amorphous silicon, on the other hand, is relatively benign.

Amorphous silicon is so common, the terms 'thin film' and 'amorphous' are often used interchangeably. As amorphous silicon is less efficient than monocrystalline and polycrystalline, amorphous modules need to have a larger surface area in order to generate the same power output. While the power output of all modules is reduced with increased shading and also by high temperatures,

Figures 3.12 and 3.12a Polycrystalline cells (left) and modules (right) can be recognised by their speckled blue appearance. Note the modules on the right are using cells which have been cut in quarters and wired together to create the required module output.

Source: John Keane

amorphous modules are impacted to a lesser extent than monocrystalline and polycrystalline modules. Multi-junction amorphous silicon, which involves multiple layers being deposited on one surface, produce higher efficiencies than single junction amorphous. They can also be more flexible and hence less fragile – an important thing to consider when modules are being used in tough environments.

Amorphous silicon modules generate higher levels of power for the first month or so of use, before dropping to, and stabilising at, a slightly lower power output. This is not a problem, so long as the manufacturer has accounted for this drop in giving the product its power rating (Wp).

PV Wiring and Junction Boxes

As pico-solar modules are designed to be placed outside, it's important to ensure that the cables and the junction box on the back of the module which link the module to the rest of the pico-solar system are able to withstand the elements. In short, they should be good quality and be sunlight, ozone, UV, and moisture resistant. They should also, of course, be long enough so that the module can easily reach and be placed outside.

Figure 3.13 A thin film module can be recognised by its dark uniform appearance with lines running along the module, generally at around 10 mm intervals. However, it is not usually possible to tell what type of thin film module it is from looking at it. Labels and datasheets need to be referred to. While many thin film modules are framed in glass, thin film can also come on flexible modules.

Source: John Keane

Figure 3.14 Flexible, Unisolar multi-junction thin film solar module.

Source: Frank Jackson

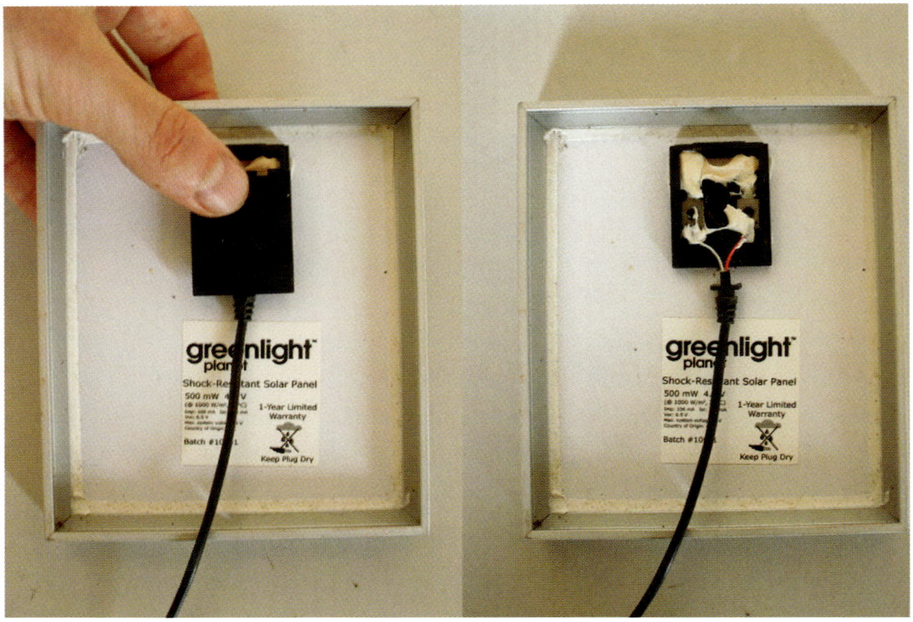

Figure 3.15 The junction box at the rear of a module projects the points where the cable and wiring connects to the module's terminals. The connections are often covered with waterproof sealant for additional protection from the elements.

Source: John Keane

Wires are connected to the positive and negative terminals at the rear of a solar module. These connection points are usually protected with some waterproof sealant and a junction box which keeps the connections out of sight.

Box 3.2 PV Module Disposal

While PV modules are generally designed to last for 20 to 30 years, they will eventually degrade, stop generating electricity and create solid waste. It is possible to recycle or reuse many of the materials, such as glass, aluminium, plastics, wiring and even some of the semi-conductor materials. The EU has recognised the importance of recycling PV modules by including them in the WEEE Directive on waste from electrical and electronic equipment (WEEE 2012/19/EU). More information on PV Module recycling can be obtained from the non-profit association PV Cycle (www.pvcycle.org).

For recycling to take place, however, facilities need to exist, be accessible and in all likelihood there needs to be some sort of economic incentive. Unfortunately, many parts of the word do not have such facilities, creating a real risk that any hazardous material contained in a module will eventually result in pollution. For this reason, it is recommended that modules destined for markets which do not have proper facilities do not contain cadmium or copper indium gallium (di) selenide.

PV Module Price Trends

The price of PV modules have fallen consistently over the past 20 years from over USD 6/W in the early 1990s to less than USD 2/W today. Figure 3.16 shows how prices have continued to decline between 2008 and 2012. Some commentators believe that the price per watt has now reached a point where, without further innovation in the technology and due to uncertainties in world markets, significant price reductions in the years to come are unlikely.

Solar PV Innovation

Over the past few years, the price fall of solar PV modules has been dramatic and largely driven by economies of scale as more and more governments have put in place incentives to promote the uptake of grid-connected solar. Many companies and research institutions are, however, carrying out research into new generation thin film solar technologies which promise to bring down the price of PV cells and modules dramatically in the future. Dye-sensitized thin-film and organic photovoltaics (OPV) for example, are two technologies being developed which promise simple production techniques and costs far lower than those seen in today's modules.

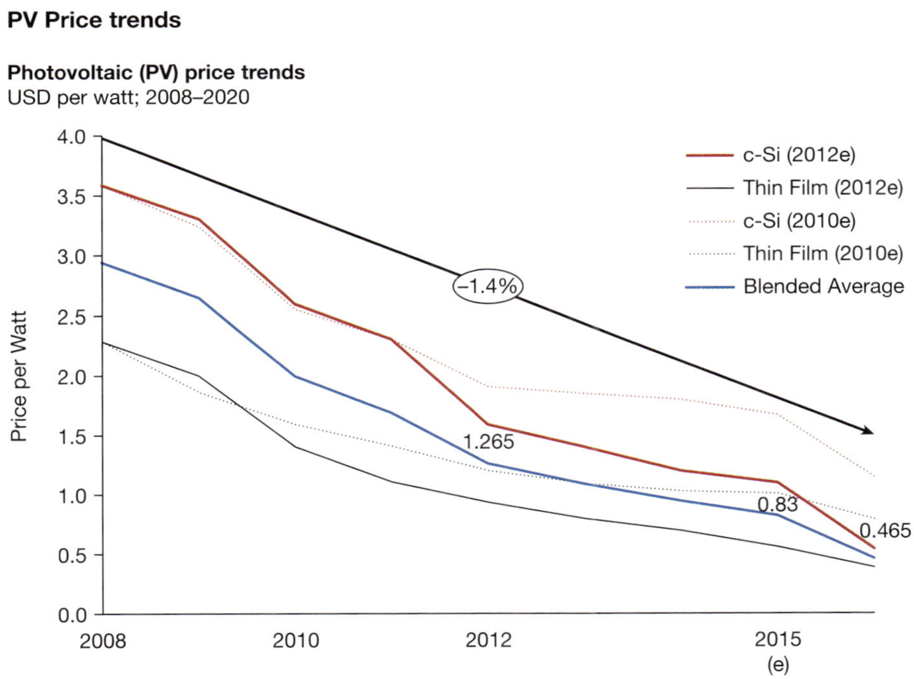

Figure 3.16 Average solar PV price per watt have fallen significantly between 2008 and 2012, with further reductions predicted.

Source: Lighting Africa Market Trend Report 2012

4

Batteries

This chapter explains how batteries work and looks at the leading battery technologies which are being used within today's pico-solar systems. It also looks at ongoing innovations in battery technology which are set to further improve product performance and lifespan.

Introduction to Batteries

Batteries come in all shapes and sizes, but they all serve the same purpose: they store electricity so that it can be used when needed. Batteries can be broadly split into two categories: rechargeable batteries (which are the type used in pico-solar systems), and disposable batteries. Disposable batteries hold a set amount of energy and can only be used once, after which they must be disposed of. Disposable batteries are not used in pico-solar systems.

Rechargeable batteries do not have to be thrown away once the energy inside has been used up. Instead, they can be charged up again, revitalising the chemicals within to store energy. One recharge is called a recharge cycle. Batteries which can be recharged a thousand times are said to have a thousand recharge cycles. The number of times a rechargeable battery can be charged up and discharged – the number of recharge cycles – varies depending on the battery type, the chemistry being used and the way the battery is recharged and discharged.

In a pico-solar system, rechargeable batteries are connected to the solar module (via a solar charge controller, discussed in the next chapter) to recharge the battery. While rechargeable batteries come in a wide range of shapes, sizes and chemistries, they all operate according to the same general principles. They are all designed to be recharged by the flow of electric current, store the energy as chemical energy and then release the energy as needed to power appliances, all the while being protected from overcharging and discharging by circuitry charge controller (see Box 4.1).

Rechargeable batteries are essentially like reservoirs of energy which can be 'topped up' with energy or 'recharged' when energy levels drop. Traditional off-grid solar systems, such as SHS, typically use lead-acid batteries to store energy. These are generally similar in shape to standard car or truck batteries, but rather than being designed to provide the short bursts of power which are needed to start a vehicle, they are designed to run appliances for extended periods of time. While some pico-solar systems use sealed lead-acid (SLA) batteries, in particular larger systems with PV modules over 5 Wp in size, it is increasingly common to see alternative battery chemistries being used, such as lithium ion – in particular

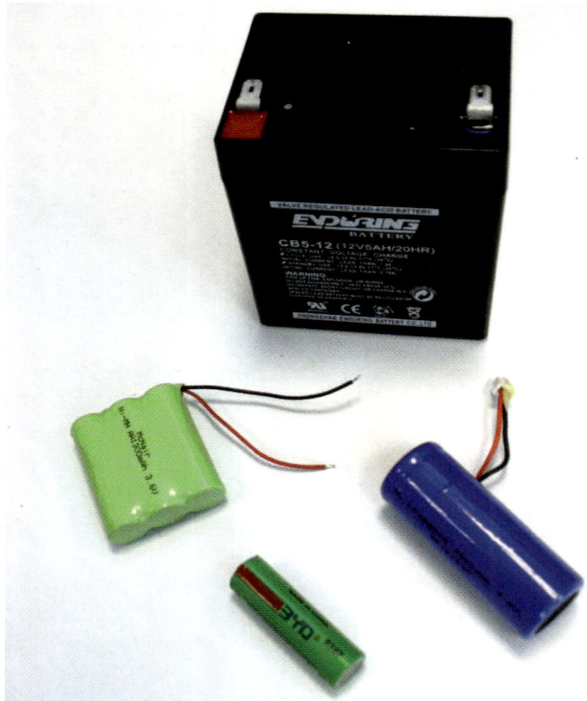

Figure 4.1 Rechargeable batteries come in a range of different shapes, sizes and chemistries as well as different voltages and energy capacities. The battery at the top is a 6 V sealed lead-acid type. The light green battery pack is 3.6 V nickel-metal hydride (NiMH) chemistry. The single green battery at the bottom is a 1.2 V NiMH. The blue battery is a 3.7 V lithium ion phosphate.

Source: John Keane

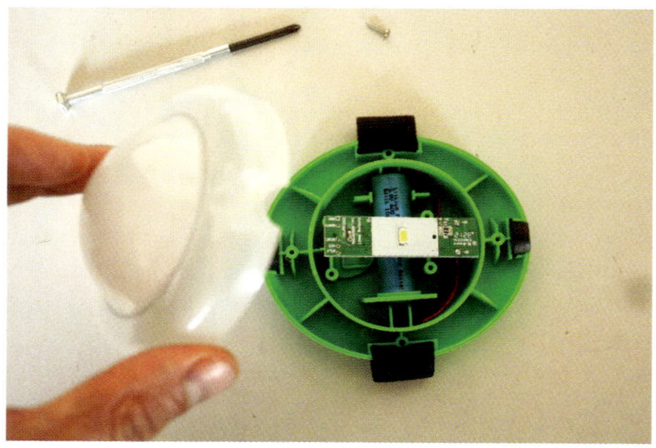

Figure 4.2 and 4.2a Pico-solar products typically incorporate the rechargeable battery within the main housing of the product. Some products enable batteries to be replaced as easily as most mobile phone batteries, whereas many other products need to be opened up with a screwdriver. Some products are designed so as to prevent everyday users from accessing the battery, so as to ensure that only certified dealers can replace a battery. This is problematic when access to a certified dealer is limited, because if the customer is unable to replace a battery, the product effectively becomes useless once the battery reaches the end of its life.

Source: John Keane

Box 4.1 Charge Controllers

A charge controller, or charge control circuitry, regulates the flow of electric current to and from a battery, preventing against over-discharging and overcharging. Over-discharging a battery is particularly damaging and should be avoided, so as to ensure a longer lifespan. Different battery chemistries require different charge control circuitry. While some larger pico-solar systems come with a separate, stand-alone charge control box which is connected to the solar module and battery, many pico-solar lanterns and battery chargers come with integrated charge control circuitry inside.

Charge controllers typically notify users on the state of charge (SoC) of the battery – this is often indicated through an interface which may be made up of LEDS indicating SoC or a more sophisticated system which tells the user how many hours the device can be used for.

Figure 4.3 Circuit board of a pico-solar lantern and phone charger which incorporates charge control circuitry and a fuse, which protects the battery from overcharging. This lantern (see below) includes a digital SoC indicator which can be clearly seen on the right side of the circuit board.

Source: John Keane

lithium iron phosphate ($LiFePO_4$) and nickel-metal hydride(NiMH). These technologies are often associated with portable appliances such as mobile phones.

Different chemistries display different properties and this has an impact on how a battery can be treated, used and maintained. The chemistries typically seen in today's pico-solar products are designed to be user-friendly, maintenance free and increasingly long lasting. They also tend to be sized differently to traditional SHS, with a lower battery capacity than traditional SHS. Pico-solar systems generally have an autonomy period of 1–2 days, versus 3–5 days in SHS, and therefore need to be recharged more often.

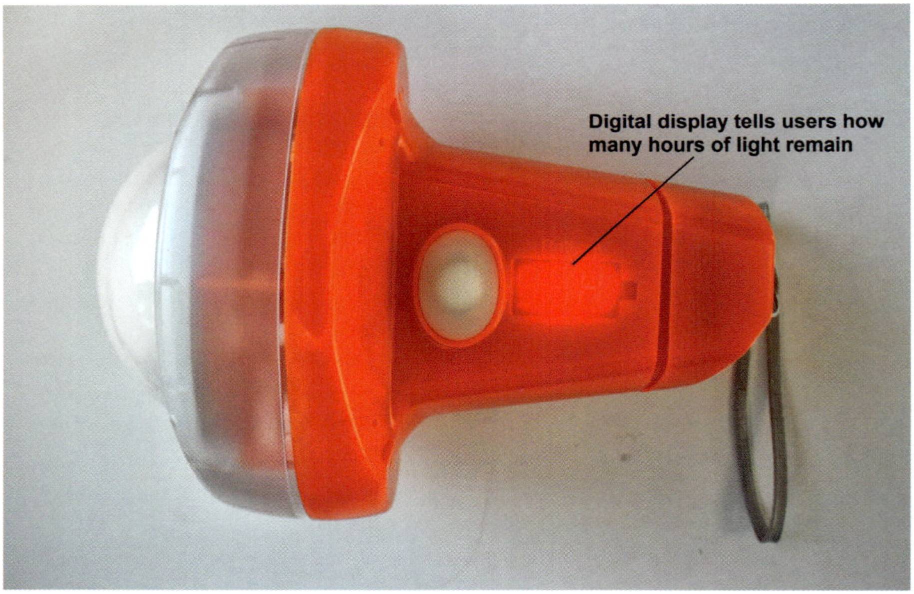

Figure 4.4 A pico-solar light with a digital display which tells the user how many hours of light remain. Other units use a simple LED indicator light to approximate the levels of power remaining. Lights like the one shown here often have multiple brightness settings. As brighter settings use more energy, users can switch to less bright settings when the SoC is low so as to extend the amount of time the light will remain switched on.

Source: John Keane

Battery C-Rates

The capacity of a battery (the amount of energy that a battery can store) is normally specified on the side of the battery in amp-hours (Ah) or milliamp-hours (mAh). This figure indicates the amount of energy that can be drawn from the battery. In theory, a battery with a capacity of 1000 mAh, for example, is capable of delivering a current of 100 mA for 10 hours, or higher currents of 200 mA for 5 hours and 400 mA for 2.5 hours, and so on. In reality, however, the rate at which a battery is discharged affects the capacity of the battery. As the rate of discharge increases (higher currents), the losses also increase, which means that the battery is capable of delivering less overall energy. Similarly, the lower the rate of discharge, the less energy is lost, effectively increasing the amount of energy a battery is able to deliver. As an example, a battery discharged at a rate of 1000 mA might provide that current for 1 hour, giving it a capacity of 1000 mAh. If the same battery is discharged at a rate of 200 mA, however, it may be able to deliver that current for more than 5 hours, meaning that the battery is capable of delivering more than 1000 mAh.

On battery datasheets, the rate of discharge is indicated as C-rate. A C-rate of C1 refers to the flow of current needed to discharge a battery fully in 1 hour. Similarly, a C-rate of C10 refers to the current flow needed to discharge a battery fully in 10 hours, and so on. The capacity of most batteries used in pico-solar systems are rated at a C-rate of C1. In practice, however, the majority of pico-solar systems used for lighting are designed to run lights for more than 1 hour, which means that the systems are designed to discharge more slowly than C1 (e.g. C10) and will deliver more current than the stated capacity.

Box 4.2 Battery Capacity

The amount of energy which can be stored in any battery is referred to as the battery's capacity. There are two ways to measure the capacity of a battery:

- amp-hours (Ah), which refers to the current that can be supplied in one hour (more accurate) Note: Pico-solar batteries often use milliamp-hours (mAh);
- watt-hours (Wh), which is the actual energy content of the battery (Ah times voltage) (less accurate).

It is more common to see amp-hours (Ah) or milliamp-hours (mAh) referred to on the side of a battery. This represents the amount of current the battery is capable of delivering in one hour. The amount of current a battery is capable of delivering does vary depending on the charge or discharge rate.

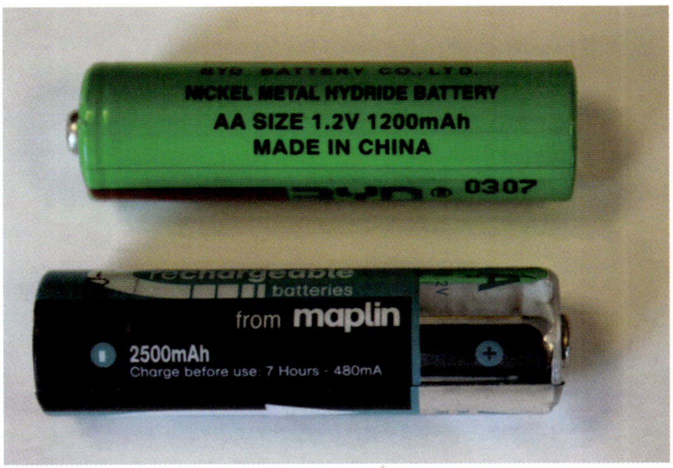

Figure 4.5 It is common to see a battery's capacity defined as mAh on the side of the battery. Batteries can be the same physical size, but have different capacities, as the example above shows – these are two AA sized rechargeable batteries, one with a capacity of 1200 mAh, the other 2500 mAh.

Source: John Keane

State of Charge (Soc)

A battery's SoC indicates the level of energy remaining within the battery. Figure 4.6 shows how the SoC rises and falls as energy is put in and taken out of the battery.

The voltage of a battery varies at different states of charge. Table 4.1 shows how the voltage of batteries with a nominal voltage of 6 V and 3.7 V, respectively, vary according to different SoC. Figure 4.7 shows a multi-meter being used to measure the voltage of a battery.

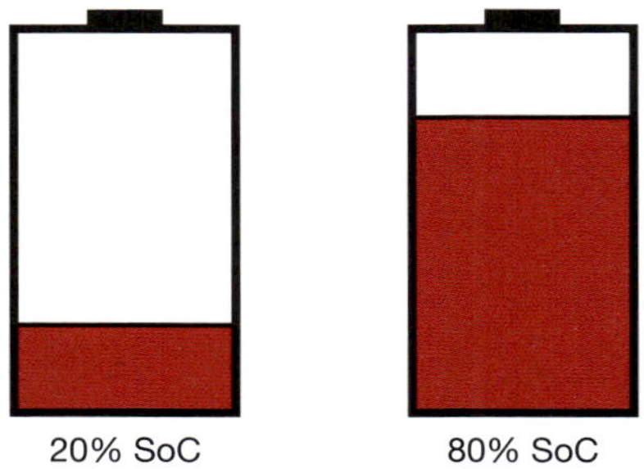

Figure 4.6 Battery SoC at 20 per cent and 80 per cent. SoC falls as the battery is used and rises again as it is recharged. When charging, the voltage of a battery rises and at full charge will actually exceed the nominal voltage (the voltage specified on the side of the battery). As the battery is discharged, the voltage drops – see Figure 4.7 below as an example. The SoC of a battery can therefore be established by measuring its voltage (see Table 4.1 top right).

Source: John Keane

Table 4.1 Battery voltage varies depending on state of discharge

Nominal 6 V	Nominal 3.7 V	SoC
6.4	4.1	100%
6.25	3.95	80%
6.10	3.85	60%
6.0	3.7	40%
5.85	3.4	20%

This example shows how the voltage of batteries with a nominal voltage of 6 V and 3.7 V respectively, vary according to different SoC.

Figure 4.7 Measuring the voltage of a battery by placing the red and black test probes of a multi-meter against the positive and negative battery terminals. This battery has a nominal voltage of 3.7 V, but due to a low SoC, the voltage reading is 3.41 V. At a full SoC, this battery measures 4.1 V. The batteries of pico-solar systems are normally enclosed within the housing of the product, such that measuring voltage is not usually possible without opening the product. SoC indicators exist on many products to inform users of the SoC status of the product.

Source: John Keane

Depth of Discharge (DoD)

Depth of discharge (DoD) can broadly be defined as the degree to which a battery is discharged of energy before it is recharged or topped up with energy. It is essentially measured in the same way as SoC. A SoC of 30 per cent means that the battery has reached a DoD of 70 per cent. As a rule, the less a battery is discharged each day, the longer its overall lifespan will be. For example, a lead-acid battery which is only discharged by 10 per cent a day will last longer than one discharged by 30 per cent every day. Table 4.2 shows how the lifespan of each battery chemistry is impacted by full DoD of 100 per cent.

The less deeply a battery is discharged, the longer its lifespan and the number of recharge cycles. This chart is intended as a rough guide. In general, a battery which is discharged every day to 30 per cent DoD will last longer than one which is discharged every day to 70 per cent DoD. Always refer to product specific datasheets.

Perhaps the key difference between traditional lead-acid batteries and newer technologies is the degree to which different battery types can be discharged. Lead-acid batteries used in solar systems typically display a DoD of 50–80 per cent, which means that it is only possible to use 50–80 per cent of the battery's actual capacity before it needs to be recharged again. Discharging the battery beyond the recommended DoD for any particular battery will damage the battery and reduce its lifespan. Figure 4.8 below provides a simple illustration of DoD.

Quality pico-solar systems have a solar charge controller, either integrated into the product's circuitry or as a separate unit, designed to protect the battery from overcharging (too much electricity) and it includes a low voltage disconnect function which is designed to protect the battery from discharging beyond safe levels. The

Table 4.2 Guide to battery lifespan (number of cycles)

Battery Chemistry	LiCoO2 (LCO)	LiFePO4 (LFP)	NiMH	SLA	NiCd
Cycle life (to 80% original capacity at 100% DoD)	500+	1000+	500–1000	2–300	300–1000

Source: adapted from *Lighting Global Technical Note*, Issue 10

The less deeply a battery is discharged, the longer its lifespan and the number of recharge cycles. This chart is intended as a rough guide. In general, a battery which is discharged every day to 30 per cent DoD will last longer than one which is discharged every day to 70 per cent DoD. Always refer to product specific datasheets.

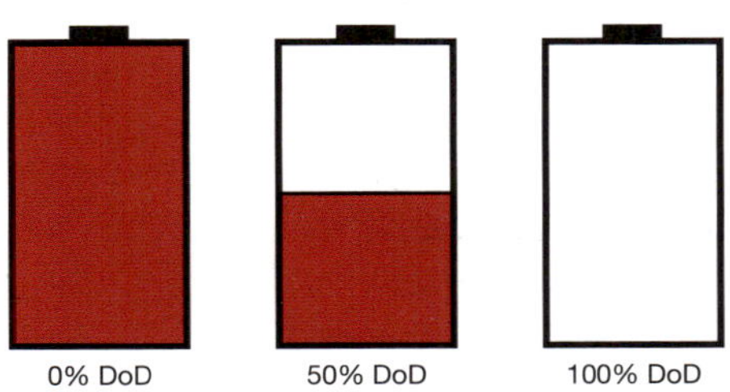

Figure 4.8 The percentage degree to which a battery is discharged is referred to as depth of discharge (DoD). In this example, the battery on the left is fully charged, the battery in the centre has 50 per cent of its energy remaining, while the third has no energy left. The degree to which batteries can be safely discharged varies according to battery chemistry.

Source: John Keane

way low voltage disconnect works is that the battery is automatically disconnected when the voltage drops to a certain level as its SoC decreases, thereby protecting the battery from levels of discharge which may cause harm.

Some chemistries, such as NiMH batteries, are generally designed so that most of their capacity can be used without causing any damage. What this means in practical terms is that if you see a sealed lead-acid battery and a NiMH, each with a capacity of 3000 mAh, the SLA battery has less usable capacity to offer. The various chemistries and their specific properties are explained in more detail in the next section.

Self-Discharge

All batteries lose charge when left unused for long periods. The rate at which batteries lose their charge varies, and is dependent on the battery chemistry and how the batteries are stored, and at what temperature. This creates problems for pico-solar products if they need to be shipped internationally for long periods and can lead to batteries being empty or near empty when the products finally reach the end customer. If left in a low SoC for long periods, there is a risk of batteries being damaged permanently, reducing their performance and functioning capacity. This is particularly a problem for lead-acid batteries. Newer lithium ion batteries suffer less from this issue, however, which is helping address this problem in the pico-solar market.

Box 4.3 Effect of Temperature on Battery Life

Batteries contain chemicals which store energy and, like all chemicals, are impacted by changes in temperature. Exposing batteries to warm temperatures will, in most cases, lead to increases in corrosion, but it is more likely that a battery will reach the end of its life before it is ended by corrosion caused by heat exposure. Pico-solar systems with integrated PV modules which require the whole system to be left out in the hot sun should be designed to protect the battery from excessive heat. In all cases, refer to relevant datasheets and manufacturers for more information on how a particular battery will perform in different conditions.

Battery Chemistries

Over the past few years, due to improvements in battery technology, pico-solar systems have gone from using batteries which last for 6 to 12 months, to batteries which last for 5 to 10 years. This is a huge difference. It has been achieved in a short space of time in part due to the fact that batteries used in pico-solar devices are in the same category of battery used in the booming portable appliance industry.

This section introduces the main battery chemistries found in today's pico-solar systems, looking at the pros and cons of each technology from performance, to overall characteristics, to price. It goes on to identify the technologies which are set to dominate the pico-solar products of the future. Each of these technologies differs in terms of lifespan, charging characteristics, the number of times

Table 4.3 Comparison of battery chemistries used in pico-solar systems

Technology	Life span (Yrs)	Cost (USD/ Wh)	Self-discharge capacity loss per month	No. of cycles	Maintenance	Recycling	Toxicity	Memory Effect	Charge Controller
Nickel-metal hydride (NiMH)	1–3	0.3	15–30%	500–1000	None	Nickel	Low	Minimal	Simple
Lithium ion (Li-ion) e.g. $LiCoO_2$	3–5	0.35	2–10%	500–1200	Keep away from heat	Cobalt		No	Complex
Lithium iron phosphate ($LiFePO_4$)	5–10	0.35	2–10%	500–2000	None			No	Complex
Sealed lead-acid (SLA)	2–4	0.25	4–8%	300–500	Keep charged	Lead Plastic	Toxic (Lead)	No	Complex
Nickel-cadmium (NiCd)	1–2	0.3	15–20%	300–1000	Discharge every 3 months	Cadmium Ferronickel	Very toxic (Not recommended)	Yes	Simple

Adapted from a range of sources – prices, recharge cycles and lifespans are estimates. Always check datasheets for product specific information.

the battery can be recharged, the type of charge control which is needed to protect the battery, levels of toxicity and efficiency and, of course, price.

Table 4.3 compares the various battery technologies:

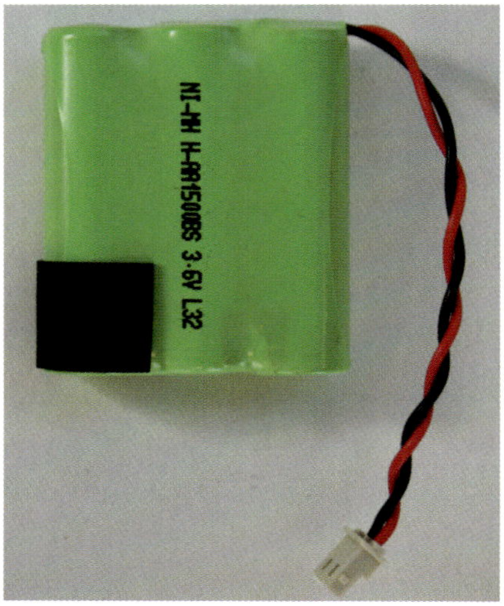

Figure 4.9 3.6 V NiMH battery pack made up of three 1.2 V NiMH batteries connected together in parallel. The wires and connector enable the battery pack to be easily connected to a pico-solar system circuit.

Source: John Keane

Nickel-Metal Hydride (NiMH)

NiMH batteries generally last for 1 to 3 years, costing approximately USD 0.3/Wh. NiMH batteries have a fairly high self-discharge rate of up to 30 per cent a month when not in use. This means that when products containing NiMH batteries are shipped internationally, they often reach their destination in a state of partial discharge such that they may need to be recharged before sale to the customer. The key advantage NiMH batteries have over NiCd (nickel-cadmium) batteries is that they do not display the same levels of toxicity and do not suffer from a memory effect to the same degree.

Lithium Ion (Li-Ion)

There are a number of different types of lithium ion battery chemistries. Two leading chemistries include lithium cobalt oxide ($LiCoO_2$), commonly found in today's laptops and mobile phones, and lithium iron-phosphate, also referred to as lithium ironphosphate ($LiFePO_4$), or LFP for short. LFP batteries have many advantages over other lithium ion chemistries (see below) and are increasingly becoming the battery of choice in pico-solar products.

Lithium ion batteries tend to have a high efficiency of between 80 and 90 per cent and will discharge by approximately 10 to 15 per cent a month when not in use. Unlike NiCd and NiMHs, li-ion batteries also last longer if they are not fully discharged before being recharged.

$LiCoO_2$ batteries have a reputation for losing charge more quickly at high temperatures and there is a risk of exploding if they are overcharged or get too hot. It is therefore important for a pico-solar product using these batteries to have a good charge controlling circuitry and also be designed so as to keep the battery in a cool place – as opposed to directly exposed to the sun's heat.

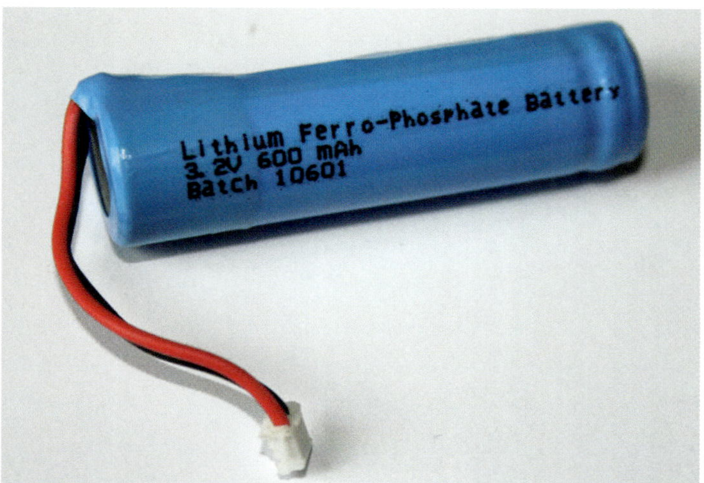

Figure 4.10 3.2 V, 600 mAh LFP battery. It is the same size as a conventional AA size battery, but at 3.2 V the voltage is higher than the usual 1.2V AA sized rechargeable batteries.

Source: John Keane

LFP batteries, on the other hand, are more stable than other li-ion chemistries. They can be stored with minimal impact on lifespan, heat has less impact on battery life and they are not known to explode if they are overcharged. LFP batteries can also achieve many recharge cycles, with some of today's products claiming up to 2000. For all of these reasons, while still more expensive than other chemistries, they are a good choice for pico-solar products and are set to dominate the sector for years to come.

Sealed Lead-Acid (SLA)

Lead-acid batteries are the oldest type of rechargeable battery and come in a number of different forms, which can broadly be categorised as 'wet cell', 'gel cell' and 'AGM' (absorbed glass matt). Wet cells are commonly used as car batteries and larger solar systems, so are not covered here. 'Gel cell' and 'AGM', however, are both SLA batteries and maintenance free, which means these batteries are sealed and do not require topping up with acid. Lead-acid batteries are cheaper than many competing chemistries such that for reasons of cost alone, they are generally well suited to larger pico-solar systems with modules of 5 W or more.

Both 'gel cell' and 'AGM' batteries, designed for use in solar systems, generally have a maximum allowable DoD of between 50 and 80 per cent, which means that you can safely use between 50 and 80 per cent of their capacity before they need to be recharged. As every model of battery is different, with its own characteristics, datasheets should be referred to in each case. Using any more than the recommended battery capacity will damage the battery, significantly reducing its lifespan. Pico-solar systems which use 'gel cell' and 'AGM' batteries therefore need a charge controller to prevent this from happening. These batteries always

Figure 4.11 6 V, 4.5 Ah SLA battery. SLA batteries are more common in larger pico-solar systems as they are less expensive than alternative battery chemistries.

Source: John Keane

need to be kept in a charged state to avoid damage and therefore need to be regularly recharged.

The main problem with AGM batteries is sulphation, which can cause them to fail. AGM batteries are at risk of sulphation if they are not kept at below full charge for long periods – several weeks or months. There is therefore a risk to retailers and end users of a battery failing within a few months if not regularly recharged. GEL batteries are relatively resistant to sulphation, but are more expensive. AGM batteries generally last for 1 to 2 years and cost around USD 0.2/Wh, whereas gel can last 3 to 4 years, costing USD 0.3/Wh. Each has an efficiency of around 85 per cent.

Lead-acid batteries are the most commonly recycled batteries in the world. Lead and plastic casings can be recovered and sold as scrap. Countries which have lead-acid battery manufacturing industries should be in a position to recycle old batteries. As an incentive, some companies offer customers a discount on replacement batteries if they return the old battery so that batteries can be recycled or disposed of responsibly. However, many parts of the world do not benefit from advanced recycling or disposal facilities. In these cases there is a real danger of toxic lead causing environmental damage to water, soil and air.

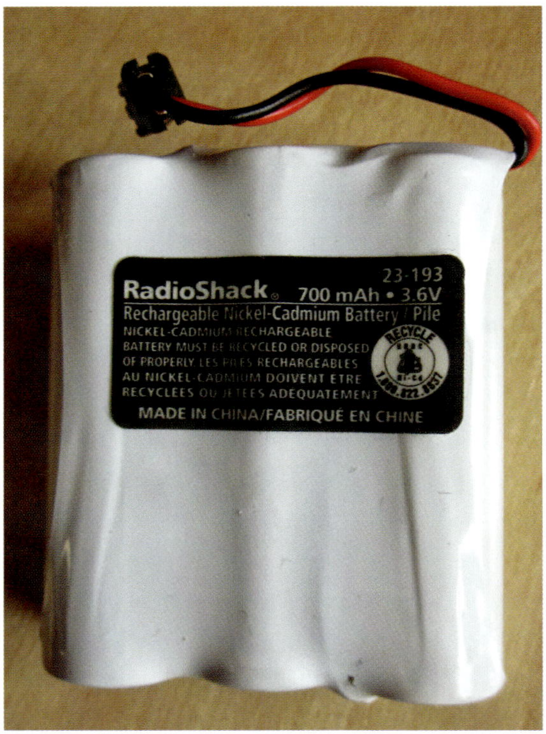

Figure 4.12 1.2 V, 600 mAh NiCd battery

Source: John Keane

Nickel-Cadmium (NiCd)

NiCd batteries generally last for 1 to 3 years and cost around USD 0.3/Wh with an efficiency of around 70 per cent. NiCd batteries are designed to be fully discharged, which means that all of the battery's capacity is available for use and they tend to tolerate around 500 cycles. NiCd batteries require a fairly simple charge controller and will discharge by approximately 10 to 15 per cent per month when not in use. NiCd batteries can also withstand long periods in a state of total discharge.

NiCd batteries do reportedly suffer, however, from the so-called memory effect, which can reduce their performance over time. 'Memory effect' is a term which was first applied to NiCd batteries to describe what happened when batteries were repeatedly being recharged after a partial discharge, so that the battery would only 'remember' to use part of the capacity used in the previous cycles. Today, the modern nickel-cadmium battery is no longer affected by cyclic memory but suffers from crystalline formation, which can reduce the battery's overall performance.

NiCd batteries are highly toxic and need to be disposed of properly to prevent the cadmium causing damage to people's health and to the environment. Where proper disposal and recycling facilities exist, there is a market for the cadmium and ferronickel that can be recovered and recycled from NiCd batteries. These batteries are not, however, well suited for use in areas which do not have developed infrastructures for safe disposal, such as rural India and Africa. As a result, NiCd batteries are not generally recommended for use in today's pico-solar products.

Box 4.4 Design Choices and Manufacturer Responsibility

Limited disposal and recycling options for batteries places more responsibility on off-grid lighting product manufacturers to use less toxic battery types in their products. SLA batteries, found in some low cost, low quality products, are of particular concern. Some of these have short lifetimes and are gaining a reputation as disposable products despite being marketed as rechargeable. SLA batteries contain large amounts of lead, and evidence suggests that the vast majority of these products are not disposed of properly.

There are two primary concerns with these types of low quality, high toxicity products. The heavy metals and toxins, when disposed of improperly, can contaminate the environment and pose a health risk to people and children who live and work in the area. Further, low quality products can contribute to market spoilage, whereby off-grid lighting products in general gain a poor reputation due to poor consumer experiences.

Manufacturers may be influenced to use NiCd and SLA batteries as these tend to be less expensive than the more environmentally benign NiMH and li-ion batteries. The design of an efficient product, however, can minimise system component sizes and help to mitigate or eliminate the added cost of NiMH and li-ion batteries. As an example, doubling the efficiency of the light source (LED, driver electronics, optics) could allow a battery pack (and battery capacity) reduction of 50 per cent, as well as a smaller solar module, without sacrificing any performance as seen by the consumer. Add to this the benefits of NiMH and li-ion batteries (high energy density, long lifetimes, small size), and the off-grid lighting product can provide good performance for the customer, long service life, and be an environmentally responsible alternative to fuel-based lighting.

Source: Lighting Global Eco Design Notes: Battery Toxicity and Eco Product Design, Issue 1, September 2012

Innovation

While batteries have developed fairly quickly as far as the pico-solar market is concerned, the truth is that the power demands of feature-filled smartphones have outpaced any improvements in battery technology. While lithium ion chemistries are likely to be at the forefront of pico-solar for the foreseeable future, it is the smartphone sector which is really driving innovation. Increasingly, power-hungry smartphones are putting pressure on researchers to develop longer lasting batteries by working with new materials, chemistries and technologies. As an example, research is reportedly being carried out to develop lightweight lithium-sulphur packs, which are likely to have a lifespan of three times that of current lithium ion batteries.

This is all good news for the pico-solar industry and there is talk of batteries in the future, even those which use existing chemistries such as $LiFePO_4$, being developed to have similar lifespans to solar PV modules themselves, thereby offering the prospect of pico-solar products with lifespans of over 20 years.

5 Lighting

One of the main uses of pico-solar systems is for lighting. This chapter introduces the reader to general lighting principles and the key terms used within the industry when measuring light output and performance. The chapter goes on to focus mainly on light-emitting diodes (LEDs) as the lighting technology which dominates the pico-solar sector. It explains what LEDs are, how they compare with other lighting technologies and how they are continuing to improve both in terms of performance and price.

Lighting Principles and Measurements

Electric lights are designed to convert electric energy into light energy, although they also produce heat energy in the process. More efficient lights minimise the amount of heat energy generated while maximising the amount of visible light emitted. The lighting industry uses a number of different methods and terms to measure the perceived brightness of visible light. The main terms are summarised below.

Luminous Flux

Luminous flux, or visible light, refers to the total amount of light emitted in all directions by a light source, as perceived by the human eye. Luminous flux is measured in lumens (lm) (see Figure 5.1).

Luminous Intensity

Luminous intensity refers to how bright a light appears to be (its level of intensity) in a particular direction. The unit of measurement is the candela (cd).

Illuminance

Illuminance refers to the degree to which a surface area is illuminated. The unit of measurement is lux. One lux equates to when a luminous flux of 1 lumen is evenly distributed over an area of 1 square metre (1 lux = 1 lm/m^2) (see Figure 5.2).

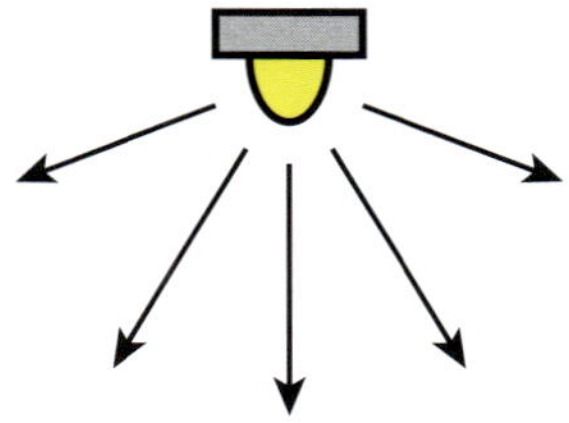

Figure 5.1 Luminous flux refers to visible light in every direction

Source: John Keane

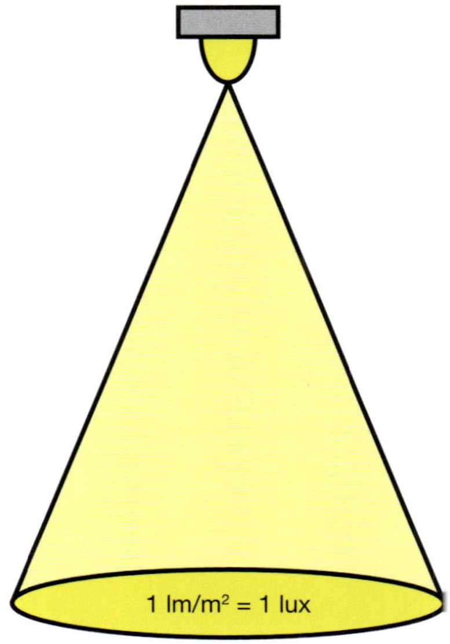

Figure 5.2 One lux equates to the even distribution of 1 lumen over an area of $1m^2$. This diagram illustrates how lux and lumen measurements relate to each other. It is important to understand the difference between lux and lumens, however. As an example, a light which provides a concentrated amount of light on a small, focused, area as opposed to providing general, all around lighting, may have a high illuminance measurement (lux) but a low luminous flux measurement (lumens).

Source: John Keane

Luminous efficacy

Luminous efficacy refers to the efficiency with which electrical power is converted into light. It is the ratio of light output to power input and is measured in lumens per watt (lm/W). For example, a traditional incandescent bulb can have a luminous efficacy of 10–15 lm/W, whereas a typical compact fluorescent bulb (CFL) is around 50–70 lm/W. LEDs can display efficiencies of over 140 lm/W and are set to become more efficient in the future.

Light-Emitting Diodes (LEDs)

LEDs are a semi-conductor source of light which have evolved over recent years into an efficient, long-lasting, lighting technology. Figure 5.3 is an illustration of what a traditional LED looks like, although shapes and sizes are evolving all the time. Figure 5.4 shows two different LED types.

Table 5.1 provides a quick, at-a-glance comparison between LEDs and more traditional compact fluorescents (CFLs) and incandescent light bulbs – the latter is only included for comparison purposes and is not a technology which features in pico-solar systems. CFLs are far more efficient than standard incandescent light bulbs and are popular across the globe as a result. Their high levels of efficiency mean that they are commonly used in SHS and are also found in some pico-solar products. As the efficiency and overall performance of LEDs have steadily improved over recent years and continue to do so, they have become competitive with CFLs. As a result of having longer lifespans, being more compact and far less fragile than CFLs, they are increasingly found in pico-solar systems in place of CFLs.

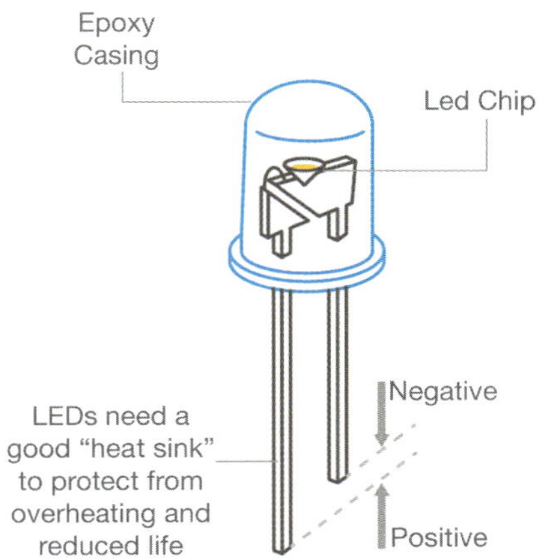

Figure 5.3 This diagram shows what a typical LED looks like, although they do come in a range of different shapes and sizes. Note: all LEDs need a good heat sink in order to control and regulate temperature.

Source: John Keane

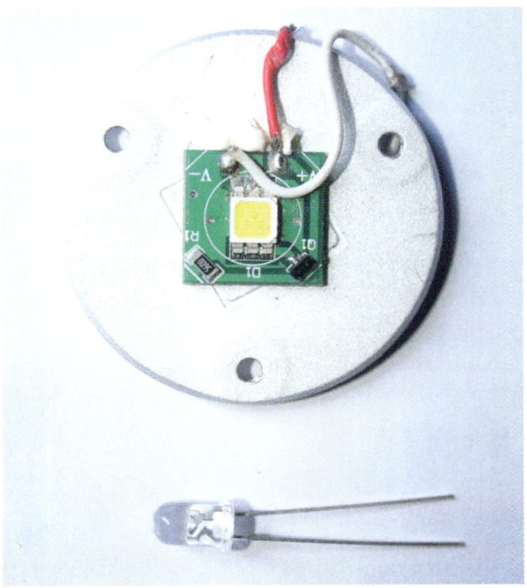

Figure 5.4 LEDs come in a range of different shapes and sizes. The LED on the left has long metal legs which act as a heat sink. The surface mount LED on the right is wired onto a printed circuit board. The round metal plate surrounding the circuit acts as a heat sink in this case.

Source: John Keane

Figure 5.5 This light has seven LEDs in the centre, surrounded by a large metallic plate which acts as a 'heat sink'. LEDs do not emit heat as infrared radiation like other light sources, so the heat must be removed from the device by conduction or convection. Heat sinks manage the operating temperatures of LEDs, prolonging their lifespan and minimising depreciation of light output over time. Without good thermal management, LEDs can display worse efficiencies and lifespans than incandescent bulbs.

Source: John Keane

Table 5.1 How LEDs compare with more traditional lighting technologies

Technology	Efficacy (Lumens/Watt)	Lifespan	Cost per 1000 lumens	Toxicity	Lumen depreciation over lifetime
Light-emitting diodes (LED)	74–144	35–50000	12	Low	30%
Compact fluorescent lamps (CFL)	63	8–10000	2–4	Contains mercury	20%
Incandescent	15	750–2000	<2	Benign	10–15%

Source: Table based on US Department of Energy figures. See http://www1.eere.energy.gov

Table 5.1 clearly shows that, all things being equal, high quality, well-designed and protected LEDs have the potential to display far greater levels of efficiency and longer life spans than CFLs. LEDs are still more expensive than CFLs, but prices are forecast to continue to fall quickly. While LEDs appear to show a greater lumen depreciation (decrease in the amount of light emitted over time), this is over a far longer lifetime. Useful life for LEDs is defined as over 70 per cent of initial light level. It is also important to note that, unlike CFLs and incandescent lights, LEDS do not generally burn out quickly, but rather depreciate slowly, over long periods. As such, the lifespan of an LED is typically measured in terms of how long it takes to depreciate by 30 per cent, but in reality can continue to emit light for much longer.

It is important to note that CFLs contain small levels of mercury and should really be recycled or disposed of as hazardous waste. As a significant proportion of the market for pico-solar products is in areas without robust waste disposal systems, this presents an issue that can create health problems for people exposed to improperly disposed of CFLs, especially if chemicals leach into soil and water.

LED development

The evolution of LEDs is still ongoing, with the technology continuing to change and evolve very quickly. At the time of writing, new generations of LED devices are becoming available every year and prices are continuing to decline rapidly. It is these developments which, together with improvements in battery technology and solar module prices, are making the rapid rise of the pico-solar industry possible.

The fact that the efficiency of LEDs is set to improve and prices are set to fall further, bodes well for the pico-solar industry as its ability to offer low power, low cost lighting. Figure 5.6 shows how recent an entrant LEDs are in the lighting sector and how rapidly luminous efficacy is improving and continuing to improve as compared to CFLs and incandescent lights.

While LEDs are undoubtedly the lighting technology of the present and the future, there are a number of important challenges and opportunities to be aware of. As with all technologies, the quality of LEDs can vary immensely, depending on the materials used during manufacture. It is therefore important to establish if the pico-solar product you are purchasing provides a warranty. Also consider asking for any test results for the LEDs any system uses which verify the quality.

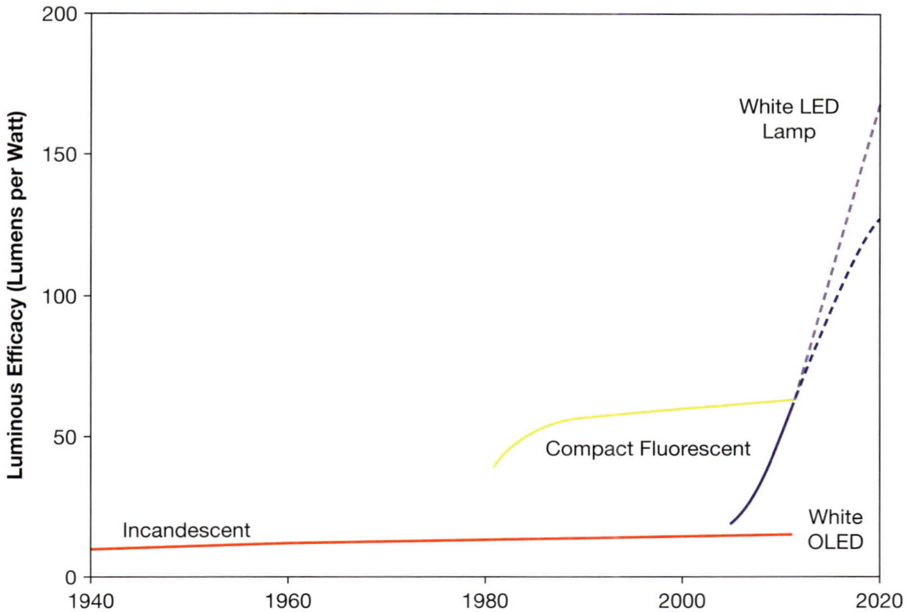

Figure 5.6 Chart showing luminous efficacy of LEDs, CFLs and incandescent bulbs over time. Note how quickly LEDs are evolving – a trend which is set to continue. Increasing the efficacy of LEDs effectively increases the amount of light output per watt. The LED efficacies in this chart include driver or ballast losses.

Source: Adapted from US DOE Solid-State Lighting Research and Development: Multi-Year Program Plan April 2012

White Light and Efficiency

While LEDs offer an efficient source of light, unlike incandescent bulbs and CFLs, LEDs are not inherently white light sources. Instead, LEDs emit nearly monochromatic light, making them highly efficient for coloured light applications such as traffic lights. While already efficient sources of light, the potential of LED technology to produce high quality white light with unprecedented energy efficiency, beyond those shown in Table 5.1, is the impetus for the intense level of research and development. The US Department of Energy's long-term research and development goal calls for cost-effective warm-white LED packages producing 224 lm/W by 2025.

Environmental Impact

LEDs are touted as being the green, energy efficient, lighting option. While this is certainly true and LEDs are designed to last for many years, all electronic products degrade eventually and need to be disposed of responsibly or recycled when they reach the end of their life. The challenge in many parts of the world, however, is the lack of infrastructure to ensure safe disposal or recycling of LEDs which can contain small amounts of heavy metal which can be toxic to the environment. It is therefore important for the industry to carefully consider how the risks of environmental damage can be minimised. More information on LED properties, design and performance can be found in technical notes available at www.lightingafrica.org.

LED Price Trends

Recent industry road mapping indicates prices for warm-white LED packages have declined by about one-third, from approximately USD 18 to USD 12 per thousand lumens (kilolumens, or klm) from 2010 to 2011. The US department of energy expects the dramatic price reductions in LEDs to continue and reach approximately USD 2/klm by 2015.

LED Lighting Innovation

The development of LEDs as efficient, cost-effective, long-lasting light sources has helped the pico-solar industry develop transformative products, especially for low income households living in areas without access to electricity. LEDs are still yet to achieve their full potential, however, and a lot of work needs to be done to further reduce costs and improve performance. Indeed, the US Department of Energy's long-term research and development goal calls for cost-effective, warm-white LED packages producing 224 lm/W by 2025. (These LEDs produced 111 lm/W in 2012.) It is fairly safe to assume, therefore, that along with projected price reductions, LEDs will continue to improve as research goes into the materials and techniques used to maximise efficiency and output as well as colour quality and thermal management. A second category of LED, the organic light-emitting diode (OLED) is also being developed. OLEDs are less developed than traditional LEDs – but they do offer the promise of more diffused lighting. As this technology develops, future pico-solar systems may well use OLEDs in some shape or form.

6
Appliances and Energy Use

The rise of small, electronic appliances, such as the smartphone, tablet computer and e-reader, all of which operate with relatively small amounts of power, has increased the number and type of electronic appliances which pico-solar systems can power. This chapter provides a broad overview of appliances which can be powered by pico-solar systems or which include an integrated pico-solar system.

The chapter also provides the reader with guidance on the kind of products which have low power consumption and provides some examples of how to calculate energy requirements and choose systems which are large enough to meet energy needs. The chapter concludes with a look at slightly larger solar systems than the average pico-solar system which can be used to run more power-hungry appliances, such as electric fences and bus shelters.

Powering Small Appliances

In order for an appliance to work in a pico-solar system, it must use low amounts of power and operate on DC at a voltage compatible with the system. While electricity grids usually deliver electrical power using AC – as it is more efficient to do so over long distances – most everyday electronic appliances, such as phones, radios and tablet computers, actually use DC. When appliances which need DC electricity are plugged into a mains socket which delivers AC electricity at 110 V or 240 V, the AC electricity runs through an adapter which converts it into DC electricity and reduces the voltage to a level compatible with the appliance. The plug for a mobile phone or an adapter of a laptop always provides details of the conversion which is taking place, specifying the AC input voltage and the output DC voltage and current. Figure 6.1 is an example of a laptop adaptor; Figure 6.2 shows a mobile phone adaptor.

As pico-solar systems generate DC electricity, no conversion from AC to DC is required in order to charge up appliances. Some pico-solar systems operate at 12 V, whereas other, smaller systems, in particular solar lanterns, operate at around 5 V. USB outlets provide 5 V and this has led to the development of many appliances operating at this voltage. See Figure 6.3.

The operating voltage of an appliance is only part of the story, however, and it is important to understand if the amount of current needed to operate the device will completely drain the battery of the pico-solar system. As an example, tablet computers and smartphones tend to operate at the same voltage and are rechargeable via a 5 V USB outlet. The tablet computer's battery may have four

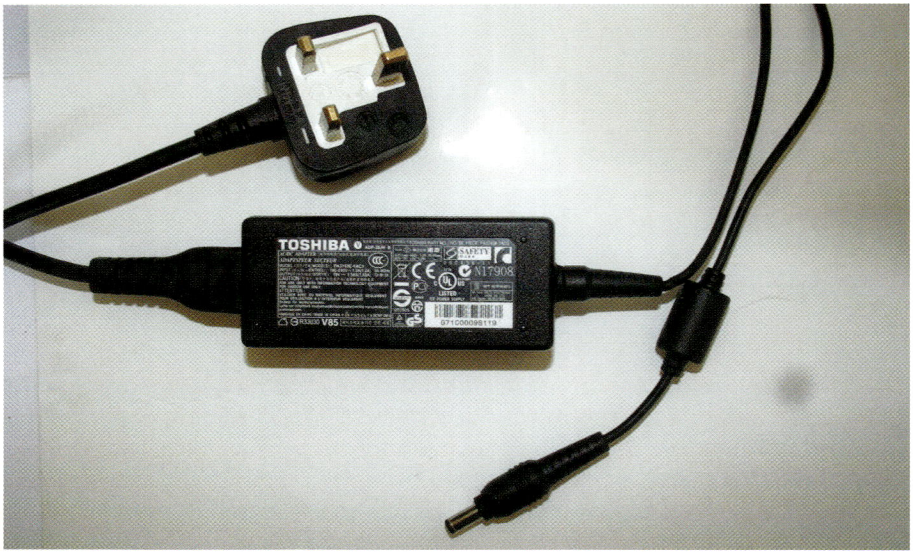

Figure 6.1 AC/DC power adaptor for a laptop. The mains plug (top left) connects to a socket to receive 240 V of AC electricity. The adaptor then converts this to DC electricity (19 V, 1.58 A) to run a laptop.

Source: John Keane

Figure 6.2 AC/DC power adaptor for a mobile phone. The label shows that this adaptor accepts an AC input of 220 V and converts to a DC output of 5 V, 500 mA.

Source: John Keane

times the capacity of the smartphone, however, and will completely drain many pico-solar system batteries which are designed only to power lights and recharge phones. It is possible to use a solar module to directly charge small appliances, but care needs to be taken to ensure the appliance is suitably protected from potential overcharging and is able to tolerate the different levels of charge a solar module generates throughout the day as the sun changes position.

Appliances which do not have an internal battery and need to be directly plugged into mains AC electricity in order to operate, such as irons and

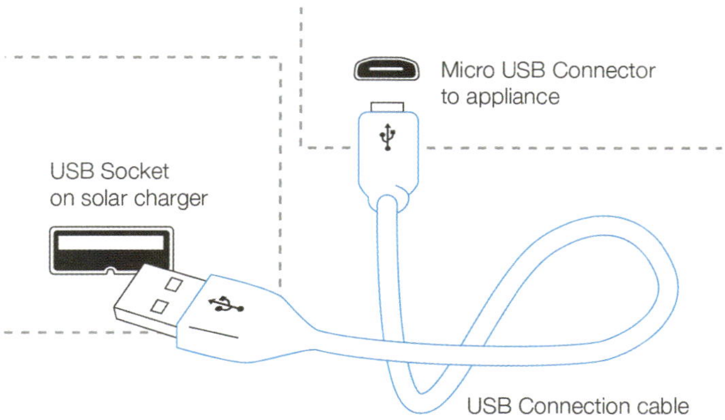

Figure 6.3 The micro USB connector has become fairly standard as a connection type used to recharge portable electronic appliances, such as smartphones, e-readers and digital cameras. USB sockets provide a 5 V power supply and usually provide a current of between 400 mA and 1500 mA. It is becoming increasingly common for pico-solar systems to incorporate a USB socket such that they are able to recharge USB powered appliances. Check the system specifications to establish the current output.

Artwork: Marianne Kernohan

hairdryers, are not compatible with pico-solar systems. It is also useful to note that appliances which are designed to generate heat require large amounts of power, which are beyond the capability of small solar systems to power. All products provide information on the levels of power they consume. Check the labels if in doubt.

Table 6.1 below provides an overview of the power and energy requirements of a selection of household appliances which operate at 5 V, indicating which are well suited to a small pico-solar lighting system and which are simply beyond its capability and require a larger system.

Table 6.1 Energy requirements of different appliances

Appliance	Power rating (Wp)	Based on pico-solar system with 5 V battery with 1500 mAh capacity	Notes
LED light (100 lumens)	1	5 hours run time	
LED light (20 lumens)	0.18	30 hours run time	
Small radio	1–10	5–15 hours (approx.) run time	Power draw varies slightly with volume
Mobile phone	2	100% charge System will fully charge up to two basic phones or one smartphone	Battery capacity typically varies between 600 mAh for rudimentary phones and 1500 mAh for smartphones
E-reader (basic e-reader device)	3	Up to 100% charge System will charge most e-readers, but may not achieve full charge	Battery size slightly larger than smartphones. E-readers with battery capacities above 1500 mAh will only partially charge
Tablet computer	10	25% charge (approx.) Will only provide partial charge for most tablets which tend to have battery capacities of around 6000mAh	This system will only provide a partial top up for tablet computers Tablets can be recharged with a larger pico-solar system with a module of 5–10 Wp and larger battery capacity above 6000 mAh

Calculating Energy Needs

Whether the system and appliances being used operate on 5 V or 12 V, it is possible to calculate how much energy a pico-solar system can produce and also to calculate the size a system needs to be in order to provide sufficient energy for any given appliance or set of appliances. Table 6.1 shows that small pico-solar lanterns which have been designed to run efficient LED lights and charge a mobile phone, will generally not be able to fully charge larger appliances with larger batteries, such as tablet computers. It is important to understand the capacity of the pico-solar system, the amount of energy needed to run or recharge appliances and the need to manage power consumption as necessary. Managing or reducing power consumption is particularly important at times when the system has not been able to receive a full solar charge.

As an example, if a small pico-solar lantern with the ability to recharge phones has an internal battery capacity of 1000 mAh, the run time of the light will depend on how much energy is stored in the battery. If the lantern is used to recharge a phone battery which has a capacity of 600 mAh, this will leave around 400 mAh of battery capacity to run the LED light. If the LED light needs 100 mA in order to operate, this will mean it will have a run time of around four hours. If the user tries to charge two phones, however, the second phone will only receive a partial charge and there will be no energy left to run the light.

In order to understand if a pico-solar system is capable of recharging an appliance, there are a few key factors to consider:

- Does the pico-solar system operate at a compatible voltage to the appliance?
- How does the battery capacity of the pico-system compare to that of the appliance? Will it charge fully? (see calculating battery section below).
- How quickly will the solar module recharge the pico-solar system? If the module is undersized in order to make the system portable, this will limit the ability of the system to be at full power (see sizing PV module section below).
- Is the charging plug/socket for the appliance physically compatible with the pico-solar system?

If the pico-solar system operates at 12 V and the appliance is designed to operate at 5 V, a voltage converter will need to be used to ensure safe delivery of power. Pico-solar products operating at 12 V increasingly have a built-in voltage converter enabling them to recharge appliances through 5 V USB outlets. In the absence of a built-in system, separate voltage converters which convert 12 V down to 5 V, delivered through a USB outlet are available (see Figure 6.4).

Daily Energy Demand

The daily energy demand for lighting, phone charging and playing a radio can be calculated by looking at the power rating of each device and the number of hours each item is used every day. Table 6.2 provides an example of a typical energy demand of a small solar system powering several appliances.

Figure 6.4 This voltage converter converts 12 V (typical voltage of car batteries) down to 5 V (typical voltage of USB outlets).

Source: John Keane

Table 6.2 Energy consumption table

Appliance (load)	Output	Watt	Hours per day	Watt-hours per day
LED light for general lighting	100 lumens	1	3	3
LED task light for study	50 lumens	0.5	4	2
Mobile phone	Charge 100%	2	2	4
Small radio	Medium volume	1	2	2
Total daily energy consumption of all appliances				11

In this example, the amount of energy needed totals 11 Wh/day, although this quickly rises if the user wants to charge more than one phone or increase their radio listening hours. The energy required must be drawn from the battery which in turn is recharged by a solar module. Both the battery and the module need to be large enough to supply enough energy and account for inefficiency losses.

The size of the module and battery capacity required to meet this energy requirement can be determined by taking a number of variables into consideration, such as the amount of solar insolation available, system inefficiencies, battery DoD and the number of autonomous days the system needs to operate without a solar charge.

Calculating Module Size and Battery Capacity

The size of the module and battery capacity can be calculated based on the energy requirement and by following the steps below.

Step 1: Calculating Energy Requirement

The daily energy demand is the amount of energy needed each day to power the appliances. This may, for example, be just one small LED light, or it could be multiple loads, such as several lights, phones and a radio. This is measured in watt-hours (Wh) or amp-hours (Ah). Wh and Ah are roughly correlated when one knows the operating voltage.

In Table 6.2, the daily energy demand is 11 Wh. If we assume the system loses approximately 20 per cent through battery and other inefficiencies, this rises to 13.2 Wh – this is the amount that needs to be inputted to the battery.

Note: battery inefficiencies vary from battery to battery and can be less than 5 per cent in lithium ion batteries and over 30 per cent in some nickel chemistries – and can increase as the battery ages.

Step 2: Calculating the Module Size

The module used in a system must be large enough such that the power it produces is sufficient to meet the daily energy requirement. It is possible to calculate the approximate size of the module needed to power a system by dividing the total energy requirement by the amount of solar energy available (insolation value) and taking inefficiencies into account.

When calculating the size of a module, an accurate solar insolation value is needed. It is good practice to establish what the daily solar insolation levels are for a site and base the size of the system on the month which has the lowest 'mean daily insolation', thereby helping to ensure that the system is able to generate sufficient power throughout the year – even in the month with the lowest average insolation. It is important to note, however, that there will be some days with lower insolation than the average, which will result in less power being generated. These variations can be taking into account by using a larger battery to store additional power.

Another method used when determining what module size is required, is to base calculations on annual average insolation levels. As many pico-solar systems are not designed with one geographic location in mind, however, insolation values will vary and designers must make many assumptions when sizing the module. If a pico-solar lantern is being designed for use in sub-Saharan Africa or India, for example, mean daily solar insolation levels will likely fall between four and seven PSH, which is a significant difference (see Table 6.3). The same lantern designed for Africa or India will struggle to operate properly, if at all, when used in northern Europe during the winter months, where insolation levels are far lower.

Another important factor to take into account is how the module is used. If the user permanently installs a module of a pico-solar system in a fixed position (recommended if possible), it is important to ensure it is well-sited so as to maximise exposure to the sun. As a rule of thumb, users should place modules in open spaces, free of shade and tilt the module towards the equator at an angle approximately equal to the latitude. At the equator, while in theory modules will perform best when positioned horizontally, a 15–20 per cent tilt to the module is generally recommended if the module is fixed so that dust is washed off by the rain.

As pico-solar systems are small and often portable with integrated modules, many users will move the system and the solar module each day, which means

Table 6.3 Average peak sun hours per day at different locations

	Average peak sun hours per day, annual and in specific months				
	Year	**Jan**	**Apr**	**Jul**	**Oct**
Nairobi, Kenya	**5.26**	6.44	5.27	3.59	5.62
Pretoria, South Africa	**5.46**	6.5	4.64	4.26	6.07
San Juan, Puerto Rico	**5.31**	4.29	6.01	6.2	4.91
Delhi, India	**4.46**	3.15	5.52	5.03	4.34
Colombo, Sri Lanka	**4.78**	4.62	5.45	4.47	4.53

Levels of insolation vary from day to day and region to region. This table shows average PSH at different locations for the year and in a selection of months throughout the year. As these are averages, it is important to note that some months display much lower PSH, particularly in winter months.

that the amount of sunlight a module is exposed to each day will vary. An educated user who knows to position the module to face the sun throughout the day will effectively track the sun as it crosses the sky in order to generate as much power as possible. Other users may be less diligent and may place a module in a location which only receives sunlight for part of the day or at less than optimal angles, such that it does not generate as much power as it could.

The appropriate module size for a system can be calculated using the daily energy requirement, the insolation available – it is common to use the average

Figure 6.5 Pico-solar module facing the overhead sun. It is important to position the module so that it faces the sun and is free of shade throughout the day. In equatorial regions, a module can be positioned virtually horizontal, at an angle of about 15–20 per cent to allow rainwater to wash the surface. If the module is not a fixed installation, users can maximise the output of the module by positioning it at an angle to face the sun in the morning, horizontally during the middle of the day and repositioning it again in the afternoon to face the sun as it moves across the sky.

Source: John Keane

number of PSH in a month (see Chapter 2) and taking inefficiencies into account. Inefficiencies include:

- a pico-solar module will not always be operating at its maximum power point (the knee on the I-V curve – see Chapter 3, Figure 3.5);
- module output is impacted by changes in temperature – the power generated falls as temperatures rise, with significant drops in the open-circuit voltage (see Chapter 3, Figure 3.7);
- losses in any circuits between the solar module and battery.

Module Output

The output in watt-hours (Wh) of a solar PV module can be roughly calculated using the following equation

$$\text{Wh} = \text{Wp} \times \text{PSH} \times 0.7$$

where PSH is the average number of peak sun hours in a month and 0.7 is an approximate performance ratio intended to take the above inefficiencies into account. This figure is an estimate and will vary in practice – it can be lower.

Module Size

The size of the module can be roughly calculated as follows

$$\text{Wp required} = \text{Wh} \div (\text{PSH} \times 0.7)$$

Example 1:
At 5 peak sun hours per day
Wp required = 13.2 Wh ÷ (5 PSH × 0.7) = 3.77 Wp

Example 2:
At 6 peak sun hours per day
Wp required = 13.2 Wh ÷ (6 PSH × 0.7) = 3.14 Wp

In these examples, a smaller module (approx. 3.14 Wp) is required when the PHS value is higher (6). The above calculations are approximations only. When solar PV modules are being selected, the module current, maximum power point, temperature coefficient and voltage are taken into consideration. Do not come to a hasty conclusion about the size of any particular module in a pico-solar system on the basis of the above calculation. Having a module with a slightly higher Wp is always a good idea.

Note: the module has to produce a voltage which is high enough to charge the battery. Table 6.4 provides a broad indication of the minimum Voc needed to charge batteries at different voltages.

Table 6.4 Module Voc

Module Voc	Nominal voltage of battery
17	12
8–10	6
5–6	3.7

Step 3 Calculating Battery Size

It is important that the battery has sufficient capacity to store the energy produced by the PV module and run the appliances. The battery voltage in pico-solar systems typically falls between 3.6 V and 12 V, depending on the type of system. Battery capacity can be calculated using the following equation:

> Battery capacity in amp-hours (Ah) = (Daily energy requirement in Wh/day × days of autonomy ÷ battery DoD)
> ÷ System voltage (V)

Using the daily energy requirement from Step 1 as an example, assuming just one day of autonomy is required and an efficient 6 volt LFP battery with 85 per cent maximum allowable DoD (0.85 when expressed as a decimal):

> Battery capacity in amp-hours (Ah) = (13.2Wh × 1 day ÷ 0.85 DoD) ÷ 6 V
> Hence battery capacity in amp-hours = 2.58 Ah

This example only allows for one day of autonomy. In reality, it would be advisable to choose a battery with a larger capacity to reduce the likelihood of having insufficient power to run the appliances.

Many pico-solar systems are designed to be as low cost as possible such that both module and battery sizes are kept to a minimum. This means that customer expectations need to be carefully managed and it needs to be clearly explained that the system will not operate optimally unless recharged fully each day. Some companies are addressing this issue by offering customers the option of purchasing modular systems which enable them to add extra PV modules and battery capacity to the system as and when they can afford it, thereby increasing the ability of the system to power more appliances for longer, as required.

Compromising PV Module Size in Portable Solar Chargers

There is a growing market for specialist, portable solar eco-chargers and devices with their own integrated module and battery which can provide power for people ‘on the go’, such as world travellers, campers and hikers who want to keep their phones, cameras and tablets recharged. The challenge some of these products face is that they are generally designed to be portable, which means that the pico-solar modules are often restricted in terms of size and therefore charging capacity. This basically means they are often undersized and it can take many days to fully recharge. Until modules are capable of producing more power from small surface areas, this problem will remain. Chargers with integrated batteries also mean that the whole device needs to be placed outside and that care needs to be taken that charging takes place in secure locations to mitigate risk of theft. More examples of appliances with integrated chargers are discussed later in this chapter.

Figure 6.6 Designers of portable pico-solar chargers often have to compromise the size of the PV module so that the product can be carried around easily. This limits the amount of power the module can generate, which means lengthy charging times for the battery. Leaving the device to be charged in the hot sun is generally not a good idea.

Source: Rick de Gaay Fortman

Portable Electronic Appliances

There is an almost endless list of electronic appliances which operate with small amounts of power and can be recharged with pico-solar systems. This has become increasingly the case in recent years with the rise of the USB outlet, which operates on 5 V. Indeed, the USB has led to the development of many new 5 V appliances, just as the traditional 12 V car battery outlet has helped create an industry around 12 V appliances. This section discusses the most common appliances pico-solar systems can be used to power. Aside from products which come with their own integrated solar module, there are essentially two ways pico-solar can be used to charge appliances: (1) indirect charging through a pico-solar charged rechargeable battery and (2) direct charging from a module.

Indirect Charging (through a Pico-Solar-Charged Rechargeable Battery)

This is the safest way to charge any appliance. The pico-solar system has its own integrated, rechargeable battery which is charged up by the PV module. The

Figure 6.7 An increasing number of pico-solar lamps incorporate a USB outlet which enables them to recharge appliances such as phones which can be recharged via USB. As the battery capacity within lanterns is limited, charging one or more phones will significantly reduce the amount of power left to run the light itself. USB outlets operate on 5 V and small lanterns like this one usually provide a current of 500 mA.

Source: John Keane

battery is then, in turn, used to charge the appliance with a steady, controlled, supply of power. An increasing number of pico-solar systems have a built-in USB outlet, making it easier to charge up USB powered devices, such as mobile phones.

Direct Charging from a Pico-Solar Module

While it is possible to connect a pico-solar module directly to an appliance such as a phone in order to charge it, unless the appliance has been specifically designed for this purpose, this is not normally recommended and can cause battery damage. This method also means that appliances can only be charged during the daytime and the device will receive varying levels of voltage and current as the levels of sunlight reaching the module vary.

Mobile Phones, Smartphones and Cameras

In recent years, mobile phone networks have spread out across the globe, far beyond the reach of our electricity grids. This means that many people can make a phone call, but can't easily charge their phone to make that call. The majority

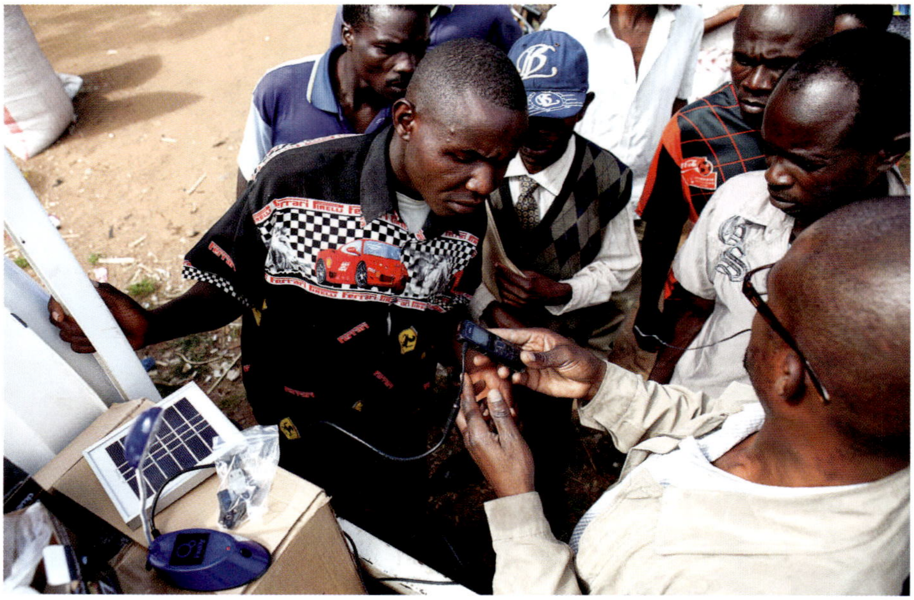

Figure 6.8 The solar module on the left is being used to charge a mobile phone directly in Zambia. While it is possible to recharge some mobile phones directly from a solar module, many phones require a constant, steady flow of electricity as opposed to the varying currents which a solar module provides. Not all phones will accept charging in this way and some battery chemistries can be damaged by charging directly from the module.

Source: © Steve Woodward

of mobile phones, smartphones and many digital cameras, have an operating voltage of 3.7 V and can increasingly be powered through USB outlets. Pico-solar systems are thus well placed to solve this problem, helping people to keep their phone on and to stay in contact with the outside world. As many people living without access to electricity are forced to pay others just to charge up their phone, pico-solar products can save low income household's money and contribute to poverty alleviation.

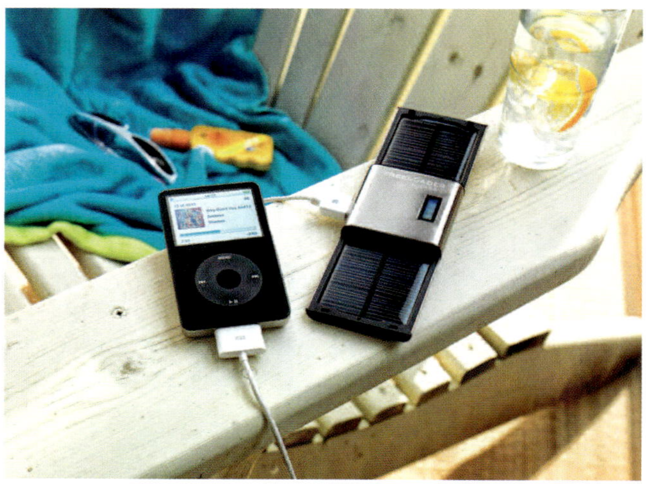

Figure 6.9 A wide range of specialised solar chargers exist with integrated solar modules and rechargeable batteries which can offer useful back-up power for portable devices such as mobile phones, cameras and MP3 players. The portable nature of these chargers often restricts the size of the solar module. This means that it takes a long time for the solar module to recharge the internal battery.

Source: Freeloader

Box 6.1 A Note on Phone Charging

First generation pico-solar products did not always charge every phone on the market. As the phone industry has now become more standardised and the new generation of pico-solar products are more advanced, this is becoming less of an issue – although the power requirements of different phone still vary slightly. It is therefore recommended to check whether a particular pico-solar product charges a particular phone to avoid disappointment.

Radios

Many radios require only a small amount of electricity in order to operate, making it possible to power non-specialist radios with pico-solar devices. The challenge, however, is that many thousands of different radio types exist. The majority of radios for sale in Africa and Asia, for example, are designed to operate with disposable batteries and only a proportion of these radios include a DC power inlet which is needed to connect a pico-solar system easily. In the absence of a DC inlet, pico-solar products can still power radios if the battery terminals can be reached. It is important to be sure that the voltage of any pico-solar system matches that of the radio, however. Some innovative companies have come up with solutions to this challenge by designing pico-solar batteries which can be placed inside the radio itself, thereby ensuring the radio operates at the correct voltage.

There are also a number of solar radios on the market, some which incorporate a solar module into the radio housing and others which are sold with a separate solar module specially designed to power the radio. As with the phone, radios with an integrated module need to be placed outside in the sun and this leads to risk of theft and exposure to the elements, such as heat, dust and rain.

Figure 6.10 Radio with an integrated solar module and rechargeable battery. This radio also has the option of being powered by mains electricity and disposable batteries for times when the solar module has not generated sufficient electricity. Designers compromise on size when integrating modules into portable appliances, limiting the amount of electricity which can be generated and increasing the amount of time it takes to recharge the battery. Integrated modules also require the whole radio to be placed outside in the sun – which is not always practical. Some solar radios come with a larger, separate module to help overcome both issues. Radios use slightly more electric current as the volume is increased. Reducing volume will therefore help reduce energy consumption.

Source: © Roberts Radio

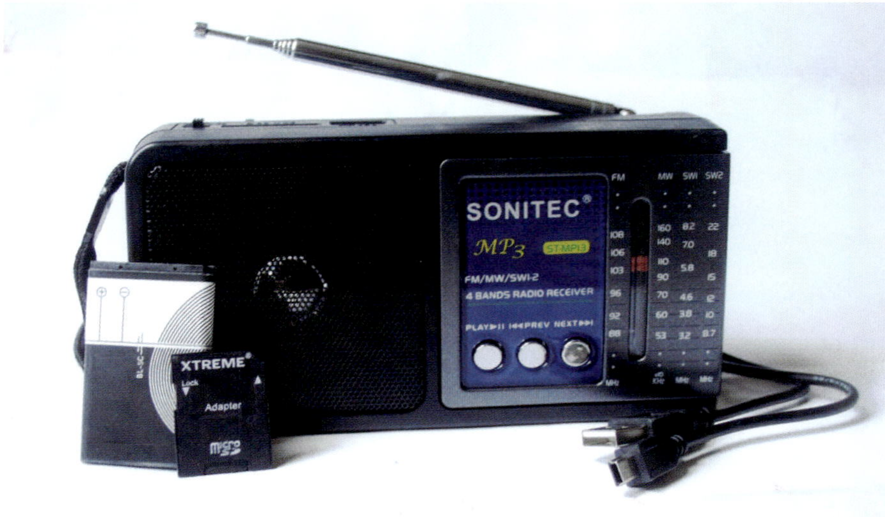

Figure 6.11 This radio available in the Kenyan market uses a 3.7 V, 800 mAh rechargeable mobile phone battery instead of traditional disposable batteries to operate. The battery can be recharged from any pico-solar system which includes a USB outlet. Products like this one reduce reliance on disposable batteries, saving customers money and reducing the number of batteries polluting rural areas without the infrastructure needed to ensure safe battery disposal. This radio uses between 50 mA and 200 mA of current to operate, which means it will operate for 4–16 hours between each recharge. The phone battery chosen to run this radio is widely available across the world, making replacement straightforward.

Another interesting feature of this radio is that it includes an MP3 player function, making it possible to insert a USB memory stick or SD card to play music or educational material. This could prove to be a very useful way for development organisations to deliver audio lessons to rural communities, enabling people to listen to programmes at times which are convenient to them.

Source: John Keane

E-readers

E-readers are well known for their electronic ink, their long battery life and claims that they can operate for many weeks before a recharge is necessary. Even e-readers need to be charged from time to time, however, the frequency of which is determined by level of use and whether the wi-fi on the device is enabled (turning on the wi-fi increases the amount of power consumed).

Pico-solar systems offer a good charging solution, especially when mains electricity is not an option. Many leading pico-solar products are capable of recharging e-readers in exactly the same way as the smartphone is being charged in Figure 6.3. It is important to note, however, that while replacement phone batteries are typically available in shops across the globe, not all e-readers have batteries which can be easily replaced, making battery replacement a potential problem after several years of use.

A number of specialised pico-solar solutions have also been developed to specifically charge e-readers, with modules and rechargeable batteries being integrated into e-reader covers. This means that the e-reader cover can be placed

Figure 6.12 This pico-solar light comfortably recharges a basic e-reader device, although this will reduce the amount of energy left in the battery to run the light. One way to maximise the amount of energy left in the pico-solar product is by recharging the e-reader during the daytime while the product is itself being solar charged. At the end of a sunny day, both the e-reader and the pico-solar product may be fully charged. A key challenge with e-readers which is common to all portable appliances, such as tablet computers, is that they will eventually need a new battery. Furthermore, not all appliances have easily accessible battery compartments and spare batteries are not always readily available.

Source: John Keane

in the sun in order to recharge. It does not need the e-reader itself to be placed in the sun. Chapter 11 includes a case study where pico-solar units are being used in a trial to recharge e-readers in schools in Ghana.

Tablet Computers

Like e-readers, tablet computers generally operate on 5 V. Unlike e-readers, however, they are far more power hungry, using batteries with capacities often four times larger than the average e-reader (6 Ah versus 1.5 Ah) which need to be regularly recharged. There are an increasing number of compact, portable pico-solar devices which can recharge tablet computer batteries, providing a useful source of back-up power on the go. These compact devices generally consist of a rechargeable battery pack with an integrated pico-solar module. The challenge with these devices is that while the battery is sized so as to fully recharge a tablet computer – for example, a 5 V battery with a capacity of 6 Ah – the pico-solar module is often undersized as the device needs to be small and portable. This essentially means that while the module can recharge the battery, it may take several days to complete a full recharge. These devices often have the ability to be recharged via USB if there is not enough time to wait for the undersized module to do its job.

Pico-solar systems which do not compromise the size of the module, as shown in Figure 6.14, recharge more quickly and are more suitable for users living without regular access to electricity and who do not need the charging system to be mobile.

Figure 6.13 A tablet computer being recharged by a portable solar powered charger. While this charger has its own internal battery with sufficient capacity (6 Ah) to fully recharge most tablet computers, its small, integrated, solar module (located on the underside) is too small for it to charge the unit quickly, which means that it will take many days to recharge its internal battery each time it is used. It can, however, be recharged more quickly from mains electricity.

Source: A-Solar

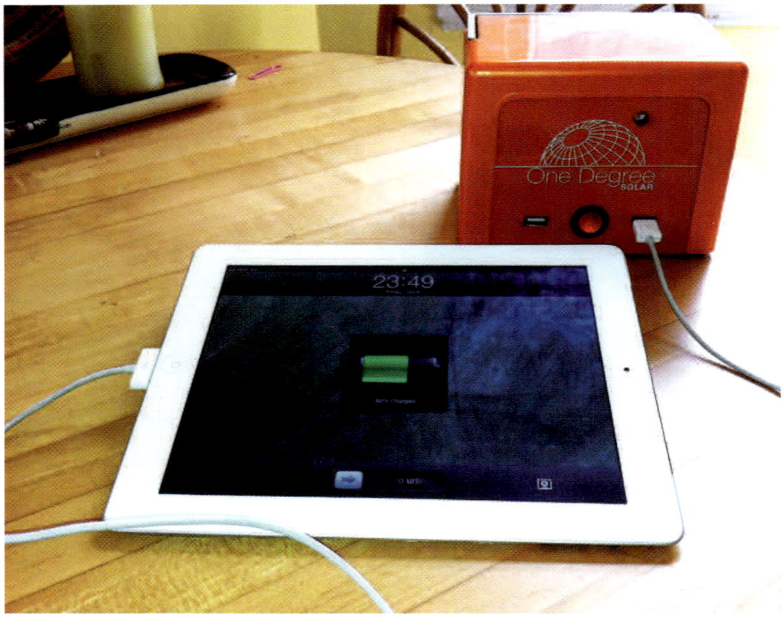

Figure 6.14 This pico-solar system uses a 5 Wp PV module and a 12 V, 7.7 Ah SLA battery, which is capable of recharging a tablet computer. In order to ensure the maximum amount of energy remains in the pico-solar battery after charging a tablet computer, it is best to recharge the tablet computer during the day so that the module is able to recharge the battery at the same time that energy is being drawn out to charge the tablet.

Source: One Degree Solar

Mobile Televisions

Televisions have traditionally been beyond the limits of what pico-solar systems can power. This is no longer the case, however, with highly efficient televisions entering the market. In 2011, for example, South Africa and Kenya saw the release of small 3.5-inch mobile televisions. This was followed by a larger 7-inch version a year later. Both devices are well within the range of pico-solar power. There is every reason to believe that the ongoing developments of efficient LEDS and batteries mean that low energy televisions with larger screens will enter the market in the future. Global LEAP, an initiative established to promote increased access to modern energy, has set up a product awards scheme to encourage this development.

Pico-Projectors

Pico-projectors are a fairly new innovation, made possible due to ongoing LED innovations. Pico-projectors currently on the market fit neatly into a pocket and generally project images of between 50 and 80 inches.

Figure 6.15 Hand-held mobile televisions which incorporate a rechargeable battery powered via a 5 V USB input have entered the market in recent years and may become increasingly popular in parts of the world without access to electricity. This 7-inch television sold through DSTV outlets can be recharged by a range of pico-solar systems and retails for around USD 75.

Source: John Keane

Figure 6.16 The pico-projector shown operates with a 3.7 V, 1600 mAh battery and will run for up to two hours before it needs to be recharged.

Source: John Keane

Lighting

Pico-solar lanterns come in a range of forms, some with separate modules, others with an integrated module. Lanterns with integrated modules need to be put outside in order to be recharged, exposing the product to potential risk of damage from water and dust. These lanterns need to be built to a high standard with an enclosure which guards against solid particles or water ingress. There is also a risk of product theft when placing it outside to charge. See Chapter 5 for more information on LED lighting.

Figure 6.17 This pico-solar lantern has an integrated solar module and rechargeable battery. The light can produce over 25 lumens for 4 hours, or operate for 8 hours at a lower light setting, from a full day's charge.

Source: © dlight design

Figure 6.18 This light, charged by a separate solar module, can produce over 300 lumens for an hour at its maximum setting, which is enough to light up a large room or classroom, or over 11 hours of light at 30 lumens, which is still enough for individual study. Note that it also includes a USB outlet, enabling it to double up as a charger for phones and other small devices.

Source: © Niwa

Pico-Solar Phone

Recognising the need and the business case for keeping mobile phones in a SoC, Kenyan mobile operator Safaricom released a phone in 2009 with an integrated solar module which quickly sold out at the time. While integrated solar modules do help recharge phones, the small surface area of a phone limits the module size and its ability to recharge the phone quickly. Integrated models also mean that the phone has to be placed outside to charge, leading to risk of product theft and exposing the device to the elements, particularly hot sunlight, and potential damage.

Laptops and Notebooks

Most laptops consume more power than tablets and operate at voltages of between 12 V and 19 V. Pico-solar systems, which operate at voltages of 12 V and less, do not generate or store sufficient levels of power in order to charge most laptops. Larger solar systems, with modules of 20 Wp and above are better

Figure 6.19 This phone comes with an integrated pico-solar module of around 0.3 Wp. Integrated modules mean that the whole device needs to be placed in the sun in order to be recharged. The limited size of the module limits its ability to recharge the phone battery quickly and struggles to recharge the battery when there is no power left. Nonetheless, the module can help top-up the internal battery and extend the time between recharges, which is useful if access to electricity is limited.

Source: © Linda Wamune

suited to charging laptops. It is also important to note that, as the operating voltage of laptops is not standardised, it is advisable to purchase a solar system which incorporates a low voltage converter to enable the output voltage of a system to be adjusted as required.

While pico-solar systems are too small to fully recharge laptops on a regular basis, there are a number of portable systems which are technically pico-solar by virtue of including a small, integrated solar module, which is capable of providing emergency recharges for laptops.

These devices typically consist of an internal battery capable of storing sufficient energy to recharge a laptop and a small, undersized, PV module, which needs several days in order to recharge the internal battery.

The modules tend to be undersized in these devices due to the requirement for portability, limiting the space available. These systems are, therefore, generally

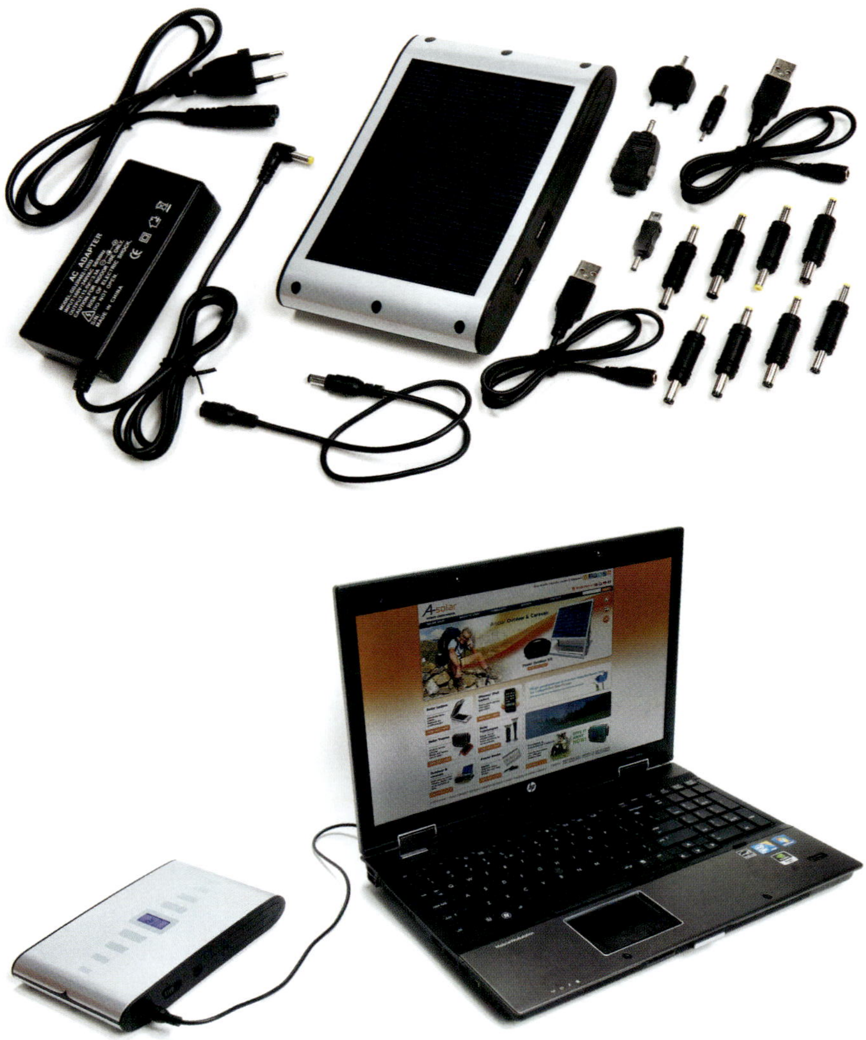

Figure 6.20 and 6.20a This device is designed to provide useful back-up power for laptops on the go. The solar module is designed to charge up an internal battery, but in reality, the module's small size means that it will take days to recharge the internal battery. For this reason, the unit comes with the option of charging through mains electricity. Users living without regular access to electricity, but who wish to regularly use and recharge a laptop, are advised to choose a system with a larger solar module.

Source: © A-Solar

more useful as a back-up power source and can be recharged more quickly via mains electricity as opposed to being recharged via the integrated module.

There are also laptops now on the market which incorporate an integrated solar module directly. Size restrictions also limit the amount of power these modules can generate, again resulting in lengthy recharge times.

Integrated modules also require a laptop to be placed in the sun; however, this puts the laptop at the mercy of the elements and at possible risk of theft. The limited power output of these modules and the need to position the whole laptop in sunlight raise questions of how practical users will find these solutions.

Solar Backpacks

There are an increasing number of backpacks which incorporate pico-solar modules on the back, charging a battery within the pack. While these are geared more towards high end markets, they do provide a convenient source of power for small appliances, for hikers who, by their very nature, spend much of their time outside.

Solar Parking Meters

It is increasingly common to see pico-solar modules perched on top of parking meters, providing the meter with a source of power. Placing a solar module on the top of the parking meter and a rechargeable battery on the inside means that parking meters do not have to be connected to the electricity grid or rely on a long-life, disposable battery in order to operate. This also means that ongoing electricity costs are reduced and less CO_2 is emitted than in other units. However, while these modules are a welcome sight to many, they risk being vandalized and they need to be kept clean if they are to operate at optimal levels. Authorities should also carefully consider the position in which meters are placed to ensure they are exposed to sufficient sunlight each day and are free of shade.

Figure 6.21 Look carefully at this image and note the integrated solar module on this laptop. Laptops with integrated pico-solar modules are appearing on the market, but one must question how practical such devices are – aside from providing nominal amounts of top-up power. The fact that the size of the module is limited to the size of the laptop means that the amount of power it can provide is also limited. Recharging should also be done with care and under supervision to avoid laptop damage (rain and dust will surely cause problems) and potential theft if placing it outside.

Source: John Keane

Figure 6.22 Solar backpacks come with an integrated solar module which is designed to recharge an internal battery to enable travellers to recharge small appliances, such as mobile phones, cameras and tablets while on the move. Systems typically include variable voltage outputs and are able to provide top-up charges for laptops. While the solar module on the backpack will recharge the battery during the day, if the backpack is being carried around, the amount of power the module generates will vary depending on whether the module is facing the sun or not. For best results, the module will need to face the sun for extended periods.

Source: Voltaic Systems

Figure 6.23 Solar charged parking meter in Sicily, Italy. Parking meters with a pico-solar module on the top recharge a battery within the housing of the meter. These systems are reportedly able to operate for over five years before a battery needs replacing, as opposed to the need to replace non-rechargeable batteries more regularly or connect the meter to the electricity grid – saving money and the environment.

Source: John Keane

Larger DC Solar Systems

While this book mainly focuses on systems with solar modules less than 10 Wp, the principles of how solar systems operate is the same for larger systems – the only major difference being that some (but not all) larger systems include an inverter in order to convert DC electricity into AC. Below are a number of examples of the different applications slightly larger DC systems are being used for.

Figure 6.24 A solar powered city bike hire station in Toronto, Canada. This example uses two 18 Wp modules to recharge a battery to run the station as an off-grid unit, despite the close proximity of mains power. In this case, using solar power means that the 80 stations around the city require no excavation or preparatory work. During the winter months, especially when snow covers modules and when insolation levels are low, the bike hire company replaces low SoC batteries as needed. Similar systems in Montreal are seasonal and can be removed in the winter months. To keep energy consumption to a minimum, the station components go into sleep mode when not in use. Similar systems can be used at bus stops to provide passengers with up-to-date service information.

Source: John Keane

Figure 6.25 and 6.25a The photo on the left shows a solar powered electric fence unit. The unit is powered by 10 Wp module with an SLA battery and charge control circuitry within the main housing. This unit can send an electric pulse along up to 30 km of electric fencing and is used to keep cows and other livestock (right) at bay on Squash Blossom Farm in Minnesota, USA.

Source: Woodstream Corporation (left), Susan Waughtal (right)

Figure 6.26 Solar module being used to recharge an SLA battery, housed in the black box below, to light up traffic signs in Rome, Italy.

Source: John Keane

Figure 6.27 A 30 Wp solar lighting system with a 12 V 24 Ah SLA battery. This system has LED strip lights and can also recharge devices such as phones, radios and small televisions through its 5 V USB and 12 V outlets. The system operates according to the same principles as smaller pico-solar lighting systems. Larger solar systems like this one often use lead-acid batteries as opposed to alternative battery chemistries such as NiMH or $LiFePO_4$, which are typically more expensive.

Source: Barefoot Power

7

Product Quality

Standards, Classifications and Tests

There are an increasing number of high quality pico-solar products available on the market, designed to function for many years. As with any industry, however, there are also many lesser quality products being sold which, at their worst, fail after just a few weeks of use. This chapter discusses product quality and provides an overview of the various standards, quality marks and product classifications which exist. It goes on to discuss different tests which can be carried out to assess a product's overall quality.

Standards and Quality Marks

Pico-solar systems are often put to use in challenging conditions, whether it is a system being used to power radio equipment on an ocean-going yacht and at risk of sea water exposure, or a system being used in a rural African household with continued exposure to dust, insects and the elements. For this reason, it is important that systems are well built to high standards which offer suitable protection to ensure the long lifespan of a system. It's no good, for example, having a product containing state-of-the-art technology if the quality of the housing holding it all together is poor or if the operating switch breaks after a month of use. The need for robust standards to protect the consumer was highlighted in a 2009 study carried out by the German organisation GTZ where 50 pico-solar lanterns were sent for laboratory tests. These tests established that many pico-solar lanterns on the market were sub-standard and displayed a number of technical problems, such as:

- poor mechanical design and workmanship;
- missing over-current protection of the LED;
- poor electrical design;
- insufficient light output;
- low quality of the LEDs;
- solar panels and batteries did not show nominal values;
- defective protection of the battery;
- defective ballast for CFLs or LEDs.

International standards exist to ensure that products meet minimum criteria and accord with international directives designed to protect both the consumer and the environment. Table 7.1 summarises some key standards and quality marks relevant to pico-solar products to look out for.

If a product meets certain standards or has won any awards, this information will invariably be advertised on the box, as these are independent verifications of a product's quality or conformity to international standards which exist to protect the customer and increase consumer trust in the product.

Product Life and Warranty

Quality products should come with a warranty of at least 12 months. It is not uncommon to see a product with different warranty periods attached to its different components, as illustrated in Table 7.2. This reflects the fact that PV modules are normally expected to last much longer than batteries.

Product Performance

All products should provide clear information which explains how it has been designed to perform. As an example, Table 7.3 explains how long a pico-solar lantern will stay on if the battery is fully charged and how long it will take to charge the battery with the solar module.

International Classifications

International customs require all goods which are imported or exported to be assigned a classification code. Most countries use the World Customs Organisation (WCO) Harmonised System (HS) to classify products. Revenue authorities use these classification codes when applying duty to imported products. When importing and exporting pico-solar systems and their various components, it is therefore important to know which classification they fall under as different categories attract different tariffs. Products which are classified under HS code 8541.40, for example, which applies to PV modules and LEDs, are duty free in Tanzania and Kenya. Table 7.4 summarises the HS codes for key pico-solar system components. How a product is classified exactly is sometimes open to a certain level of interpretation.

Product Testing

It is important that any product entering the market has undergone thorough testing to verify it meets minimum quality and performance standards. This section provides an overview of the standards being adopted for off-grid lighting products and the types of laboratory tests products are subjected to. It also provides a checklist, to help readers review and assess a product's overall quality, performance and functionality and examples of how to carry out some simplified laboratory style tests at home.

Table 7.1 Standards and quality marks

Standard	Overview
CE	The CE mark is a mandatory marking for goods sold in the European market which the manufacturer must add to a product to confirm that it complies with EU safety, health and environmental protection legislation. While it is not a mark of quality and is only relevant for goods sold in the EU, if the product you are thinking of buying does not have the CE mark, it is important to find out why not, as it likely means that the product could not be sold in Europe and that it does not meet some basic requirements which are designed to protect the consumer. This mark indicates compliance with RoHS (see below).
(Restriction of Hazardous Substances) RoHS	RoHs is a directive introduced by the EU to restrict the use of hazardous substances, such as lead, mercury and cadmium in the manufacture of electrical equipment. Interestingly, batteries are not covered by this directive and thin film PV solar modules benefit from an exception such that they are allowed to contain cadmium telluride (CdTe). Nonetheless, if the product in your hands does not have the RoHS mark on it, it may well not qualify and you should think twice before buying it if you support restricting the use of hazardous substances and the product is intended for use in areas which do not have well developed disposal and recycling facilities. More information on RoHS can be found in Lighting Global Eco Design Note 3, www.lightingafrica.org
IP Code (Ingress Protection Rating) IP 6 7	IP stands for Ingress Protection and relates to the degree to which electronic equipment is able to withstand water, dust and other solids from entering the unit. It is a European standard. An IP rating normally consists of two digits, such as IP54, with the scale generally running from IP 00 to IP 68. The first digit relates to the degree to which the appliance is protected against the entry of solids, the second relates to the degree to which the appliance is protected against the entry of water. The higher the number, the higher the level of protection. In the case of pico-solar products, IP54 is a good standard to aim for – the 5 confirming a product's ability to keep dust out and the 4 confirming ability to resist sprayed water, such as rain.
Lighting Global	The Lighting Global quality assurance programme, established as part of a joint IFC/World Bank initiative, has developed a set of minimum quality standards and recommended performance standards for pico-solar lighting products. These standards cover a range of factors, from overall build quality, to run time and lumen maintenance of LEDs. More information can be obtained from www.lightingafrica.org
National Standards e.g. KEBS KEBS	An increasing number of countries across the world are introducing their own agencies to ensure that products sold in the local market meet certain quality standards. In Kenya, for example, products need to be certified by the Kenyan Bureau of Standards (KEBS). Products which pass the quality tests will display the relevant logo.

Table 7.2 Example of how warranty periods can differ

Component	Expected life	Warranty period
PV Module	20 years	10 years
Battery	5 years	2 years
LED	2000 hrs @ ≥ 95%	2 years

Different warranty periods for different system components can be confusing for customers who may not know which part of the system is defective. As with all warranties, the key challenge is ensuring that customers know how to return a product and benefit from the warranty.

Table 7.3 Pico-solar lantern performance specifications

Product performance	
Solar charge time (in full sun)	5 hours
LED run time (on full charge)	5 hours at 90 lumens (bright setting)
	60 hours at 7 lumens (low setting)

Table 7.4 HS codes relevant to pico-solar systems

Item	HS code
PV modules	8541.40 (Photosensitive semiconductor devices)
LEDs	8541.40 (Light-emitting diodes)
Rechargeable batteries	8507
• Lithium Ion (Li-ion)	8507.60
• Nickel-metal hydride (NiMH)	8507.50
• Nickel-cadmium (NiCd)	8507.30
• Lead-acid (Maintenance Free)	8507.20
Solar lanterns	8513.10.90

The Eyeball Test

Before spending time and money subjecting a product to rigorous quality and performance tests, in addition to establishing whether a product complies with internationally recognised standards and comes with a warranty, it is important to simply try a product out to see how it feels and if it is able to satisfactorily carry out the desired functions. If possible, try using the product for an extended period of time or ask other users for their experience of use. Table 7.5 provides a short checklist summary of some important things to look for.

Table 7.5 Checklist – basic things to look for when assessing a product

Build quality	**Comment**
How strong is the product? Does it look like it would still work if dropped? Is the PV module well protected in a strong frame?	Product should be able to withstand a one metre drop. It should still be intact and function.
Does the product have lots of openings or gaps where dust or water may be able to reach and damage the circuitry? (See IP ratings section above)	A simple test is to put the module out in the rain (or in the shower) – does the product still function properly afterwards?
Are the cables strong and UV resistant?	Check for labelling. Cables may state if they are UV resistant.
Is the switch robust? Does it look like it will last being switched on and off over and over again – thousands of times?	One laboratory test is to turn the switch on and off 1000 times and see if it still functions.
Do the connections (jacks, input sockets etc.) look robust? Will they withstand rough use over long periods? Is the junction box well built?	USB connections can be quite fragile and users often force USB leads in the wrong way.
Does the internal circuitry look strong or does it fall apart as soon as it is touched? Are solders bright and shiny or dull and grey? They should be shiny.	Poor quality products use thin, low grade wire which breaks too easily. Grey solders indicate a cold solder joint (soldering iron was not hot enough), which means the connection may fail in time, be intermittent or result in inefficiencies in the circuit.
Quality marks and warranties – does the product display standard marks, e.g. CE and RoHs, and does it come with a warranty?	Leading products are offering warranties of two years. Also look out for any awards the product may have won. Ask supplier if in doubt.
Performance	
Functionality – does the product perform all the functions described on the box? i.e. – Light output – is the light as bright as expected? – Does the unit charge a selection of different phones/ appliances?	Guidance on how to test light output using a lux meter is provided in the next section. While phones tend to have similar power ratings, slight variances can result in phones not charging with certain pico-solar chargers.
Specifications – does the unit clearly explain: light output (lumens/lux); battery capacity (Ah); module ratings (Wp, Voc, Isc, Imp, Vmp, STC)	Good products provide comprehensive information. Does the product provide enough light and power?
Run time – how long does the product run an appliance (e.g. light) when: – Fully charged – Charged from empty with one full day of sun	To test run time of a light, all that is needed is a stopwatch to time how long the light remains illuminated. More guidance for testing run time with light output is provided in next section.
Length of cable (if the product comes with a separate module)	How long is the cable? Will it easily reach a roof?

Table 7.5 Continued

Ease of use and repair	
Is the product easy to use or are instructions confusing?	Instructions should explain how to use and care for the product – not add to the confusion!
Does the product have a SoC indicator which explains energy levels?	It's important to know how much energy the product has at any one time.
Does the product have charge control which protects the battery from overcharge and discharge?	This should be explained on the packaging of the product.
Is the battery compartment easy to access? Will it be easy to replace the battery when it needs a new one?	An example of battery replacement is provided in Chapter 8.
When was the product manufactured?	
Is the date of manufacture on the product or the box? It is important to know if the product has been sitting on the shelf for a long time as batteries do degrade over time if they remain dormant.	Check the battery SoC. If it is not close to 100% the battery may have degraded over time. Always recharge immediately.

Home Laboratory Tests

While is possible to get a general feel for a product's overall performance and quality just by looking at it and using it, there are additional tests which can be carried out at home with the use of fairly inexpensive equipment to see how a product performs. This section explains how to:

- test light output;
- test the autonomous run time of pico-solar lights.

Test Light Output

Objective: To test the illuminance of a product: lux over a 0.1m^2 surface.
Equipment needed: lux meter, large sheet of paper, marker pen.

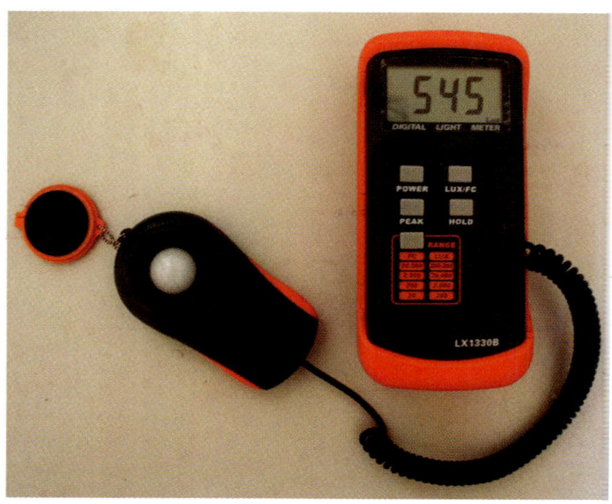

Figure 7.1 To measure illuminance, position the photo detector of the lux meter (left) so that it faces the light source. The illuminance value will then appear on the digital display (right).

Source: John Keane

Table 7.6 Test light output

Method

Position light source to be tested opposite a flat surface with an area of 0.1m^2.
In this case a study light is used as an example.

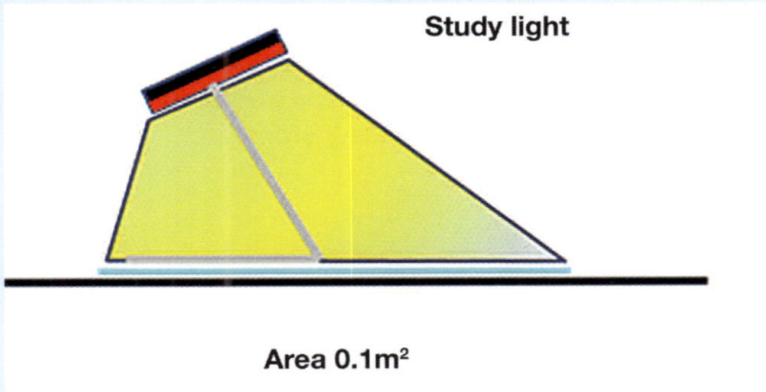

Mark the surface in a grid pattern. Measure illuminance by placing the lux meter at each point on the grid and recording the lux value at that point.

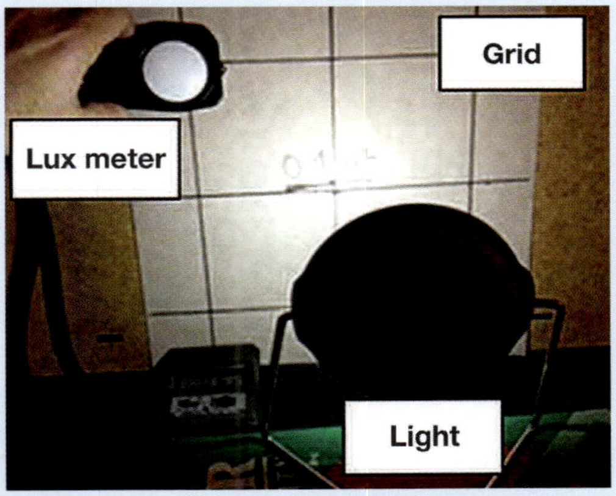

The results can be used to create a surface plot showing the distribution of illumination across the area. Conclusions as to whether the light output meets the required levels can be made.

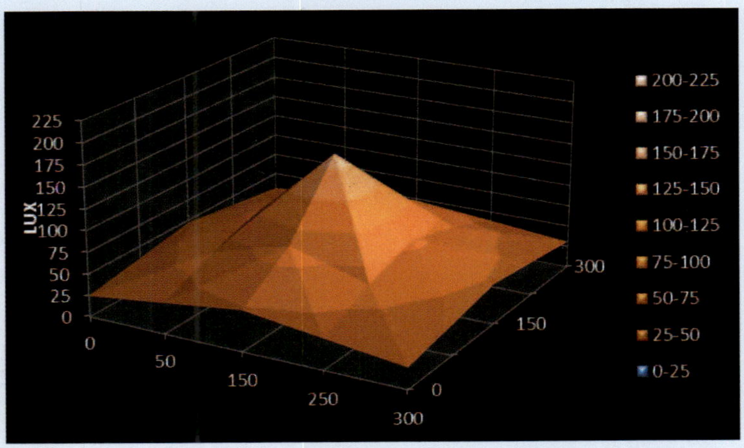

Autonomous Run Time Test

Autonomous run time means the time duration in which a product can power an appliance or light, whether from a full charge or from a full day's solar charge (depending on what run time is being tested).

Objective: To test the autonomous run time of a light.
Equipment needed: stopwatch, lux meter, integrating sphere.

> **Box 7.1 Integrating Sphere**
>
> An integrating sphere can be made from a 500 mm diameter polystyrene hollow sphere, coating the inside with Lumina high reflective paint. The outside surface can be covered in reflective foil to minimise light loss as well as light pollution from outside. The sphere should have two viewing ports; a lux meter is placed in one port and the larger diameter port is located at 90^0 to the measuring port with an internal baffle to prevent light from the input port shining directly onto the lux meter port. This integrating sphere is best suited to side-by-side comparisons between different lights rather than highly accurate absolute readings of output. See Table 7.7 for a photo of an integrating sphere.

Table 7.7 Autonomous run time test

Method	
Ensure product to be tested is fully charged or has received a full day of solar charging from empty. Check SoC using relevant indicator. Once the product has a full SoC, switch on light and start the timer.	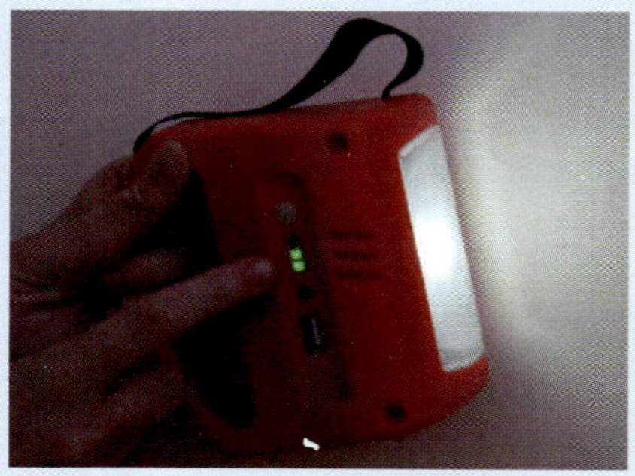
Measure the light output periodically (e.g. every 30 mins) using a lux meter and integrating sphere until light turns off (or falls below required level e.g. 50% of original). See Box 7.1 for instructions on how to build an integrating sphere. The integrating sphere blocks out all light other than the light being tested. It diffuses this light uniformly and enables optical power measurements to be taken. Plot light output results (lumens) over time.	

Table 7.7 Continued

The chart shown plots results for three identical lights, tested for comparative purposes.

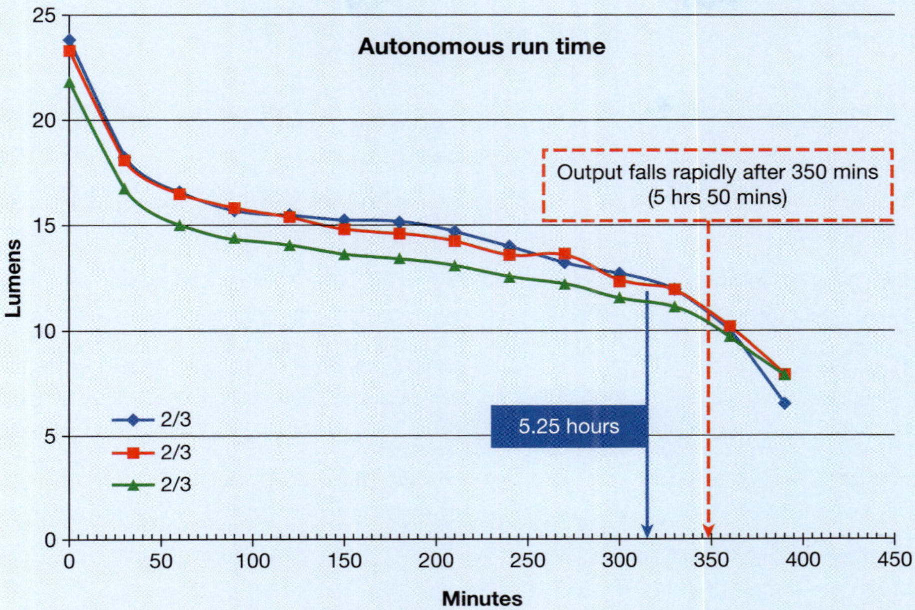

Laboratory Tests

For more in-depth quality and performance tests, products should be sent to a laboratory to ensure compliance with standards and performance targets. As a minimum, products should be tested for:

- durability and workmanship, e.g. overall build quality of wiring and circuitry;
- lighting performance and durability, e.g. lumen maintenance over time and light distribution;
- battery performance, e.g. battery capacity and efficiency; performance at different temperatures;
- solar module, e.g. module I-V curve performance;
- run time, e.g. on full battery, and battery charged by one solar day.

Following extensive tests, the German Fraunhofer Institute Test Laboratory has made recommendations which should be followed to help ensure product quality. These are reproduced as Table 7.8. Lighting Global has also developed a product test procedure document which can be downloaded here: http://www.lighting global.org/

Table 7.8 A wide number of tests should be conducted to ensure that a product meets minimum standards

Criteria	Issue	Remarks
Basic components	• Plug • Load controller	Plug for external devices, such as radio, mobile etc.
Performance	Daily burn time (duty cycle)	3 hours of light per one day recharge
Performance	Maximum run time	6 hours of light with full battery
Brightness	Enough to read: to illuminate a room: clearly brighter than regular oil lamp	Illumination: min. 300 lux (on a table, for example) Lumen: min. 150 lumen (oil lamp)
Manual	Manual must be provided in English language (comics and language of user better)	• operation • maintenance • do's and don'ts
Guarantee	Producer must issue a guarantee on performance and lifetime of lantern and its components	2 years
Ambient condition	Lamp must cope with typical ambient conditions and ensure required performance	• bright sun • dust • insects • water • humidity • temperature: 5 to + 45°C
Lifetime	Light	At least 1000 switches and operation of 2200 hours? No blackening of more than 10%
Lifetime	Battery	Load cycle 750 (2 years with 1 cycle per day) → performance requirements must be fulfilled Battery must be stored fully charged free of damage: 20°C → 6 months: 30 °C → 4 months: 40 °C → 2 months
Lifetime	Panel	PV panel must be scratch resistant PV panel must show certified performance of 90% after 5 years
Durability	Switches, plugs and all other moving pares	Must withstand 1000 cycles and function

Table 7.8 Continued

Criteria	Issue	Remarks
Energy efficiency	Luminous efficacy	The luminous efficacy of the lamp inclusive of the power requirement of the inverter, must be either: (a) greater than 30 lumens/W with any reflectors, lenses, covers or grids (if used) in place; or (b) greater than 35 lumens/W without reflectors, lenses. etc in place.
Labeling	Every lamp must show basic information	• main tech. details (for light. plug, etc.) • manufacturer • serial no. • model no.
Energy efficiency	Energy losses with no operation	No electricity losses shown with light switched off.
System protection	Components need electrical protection Charge control	• battery must be protected against deep discharge (active charge controller for lead acid and lithium ion batteries) • battery must be protected against overcharge • PV panel must be protected against reverse polarity
Shipping	Suitable packing	Vibration resistant

Source: What Difference can a PicoPV System Make? Early Findings on Small Photovoltaic Systems – An Emerging Low-Cost Energy Technology for Developing Countries (2010)

This table shows the wide number of tests which should be conducted to ensure that a product meets minimum standards which are designed to protect the consumer. Note that while inverters are mentioned in this table, inverters are not used in pico-solar systems which operate with DC only.

Box 7.2 Quality Assurance

There is a need to ensure that pico-solar lighting products, destined for markets such as rural Africa, are of a high quality. Lighting Africa summarises the challenge as follows: ‘A product performance dilemma is impacting the off-grid lighting market. There is a common disparity between stated product performance and actual performance, and without clear information, distributors and end consumers are unable to make informed purchasing decisions’.

To combat these sorts of issues, a quality assurance system was established which includes:

- testing methodology which evaluates product performance;
- product performance verification system which evaluates truth in advertising: in short, confirming whether a product does what it claims it can do;

- advisory services to help manufactures develop higher quality products;
- access to a test laboratory, enabling manufactures to accurately measure and, where appropriate, improve product performance.

As a result of this initiative, companies can submit their products for testing, using the Lighting Africa Quality Test Methodology (QTM), to establish if they meet minimum standards and performance targets. An International Electrotechnical Commission (IEC) standard has now been developed based on this QTM, such that passing the QTM is increasingly becoming the global requirement for governments and donors to provide support to pico-solar lighting products and companies.

8

Using, Maintaining and Repairing Pico-Solar Systems

This chapter provides guidance for distributors and retailers on how to sell pico-solar products responsibly. It explains how to ensure that products sold to the customer are fully functional and end-users understand how to use and maintain products. It also outlines the importance of after-sales service and repair, and the support that should be requested from product manufactures. The chapter also provides examples of how to troubleshoot for problems and carry out basic product repairs.

Selling Pico-Solar Responsibly

Pico-solar systems are generally designed to be fairly easy to use and relatively maintenance free. It is important, however, to ensure that:

- customers receive products, especially those which may have been stuck in shipping for several months, in a fully charged and operational state;
- customers understand how to use and maintain products as well as the benefits of pico-solar systems;
- customers receive strong levels of after-sales service and know where they can purchase spare parts and have their products repaired.

Ensuring Customers Receive Fully Charged Products

Minimise Time in Transit

As all batteries self-discharge over time, it is important, especially when products are being shipped internationally, to take steps to minimise the time it takes for products to travel from the factory to the point of sale. Transport times can be reduced by:

- air freighting products – this is usually much quicker than sea freight, but it is also more expensive and less environmentally friendly;
- minimising time in customs – ensuring all import and export documentation is in order so as to minimise the time it takes for products to move through customs. This means ensuring all pre-shipment requirements are completed,

the product HS classification code is clear, all import taxes and tariffs are paid swiftly. For inexperienced importers, it is advisable to use the services of an experienced clearance agent.

It is also important to keep products in cool, dry and clean conditions so as to help ensure the products sold are in perfect condition.

Check Battery State of Charge (SoC)

On arrival from the supplier, it is important to check the SoC of product batteries. Most products have simple indicators which indicate this. If the battery is below 100 per cent, the product should be recharged either by placing the unit out in the sun, or in some cases, there may be an option to use a DC adaptor to recharge the unit through mains electricity (check with the supplier).

Educating Customers

It is important to ensure that customers fully understand how to use and maintain products so as to help ensure they get the maximum performance from the system. It is also useful to be able to clearly explain the benefits of pico-solar products.

Day-to-Day Use and Maintenance

If the customer does not understand how to use and maintain a system or what its limitations are, they may come back with questions or believe that the system is not working properly. It is therefore important that customers understand the basic principles of how solar power works, as many will never have used or owned a system themselves. As a minimum, customers should understand:

- that solar module is recharged by the sun's light (not heat) and should face the sun;
- the module should be kept clean, free of shadows and will take longer to charge when cloudy;
- there is a rechargeable battery inside the system which stores energy;
- the time it takes to fully charge the system's battery and how to check the SoC;
- the limitations of the system – what the system can and cannot power;
- how to take care of systems – nothing is indestructible or theft proof!

A good product should have clear user instructions with plenty of images (see Figure 8.1 for an example). Also ask the supplier if they have a list of Frequently Asked Questions (FAQs) which can help address common and product specific customer queries. Table 8.1 lists some common questions, however retailers should expect many more questions, the answers to which will vary from product to product.

Table 8.1 Sample of frequently asked questions

FAQs	Comment
How do solar products work?	Solar modules convert sunlight (not heat) into electricity which charges up a battery which can power lights and appliances.
Does the system work when it is cloudy?	The product will still charge when there is cloud cover, however it will charge much more slowly than when there is bright sunshine.
Does the system work in the shade?	If the solar module is in the shade, it will generate less electric current and charging will take much longer. For best results, keep the module away from shaded areas.
Can I connect an inverter to the system?	No – inverters are only useful in larger solar systems. There is not enough electricity in this system for an inverter.
What do I do if the product stops working? Where can I have the product repaired?	This is a common question. Retailers should work with suppliers to ensure products can be replaced if they fail within the warranty period or repaired locally as needed, which means spare parts need to be made available.
Can I use the system to play my television?	The answer to this question is probably no! Most pico-solar systems are too small to power normal televisions. The exception is if the customer owns a mobile, USB-powered, television. In this case, if the system can power a phone, it may be able to power the television – but may use up more power than the phone.

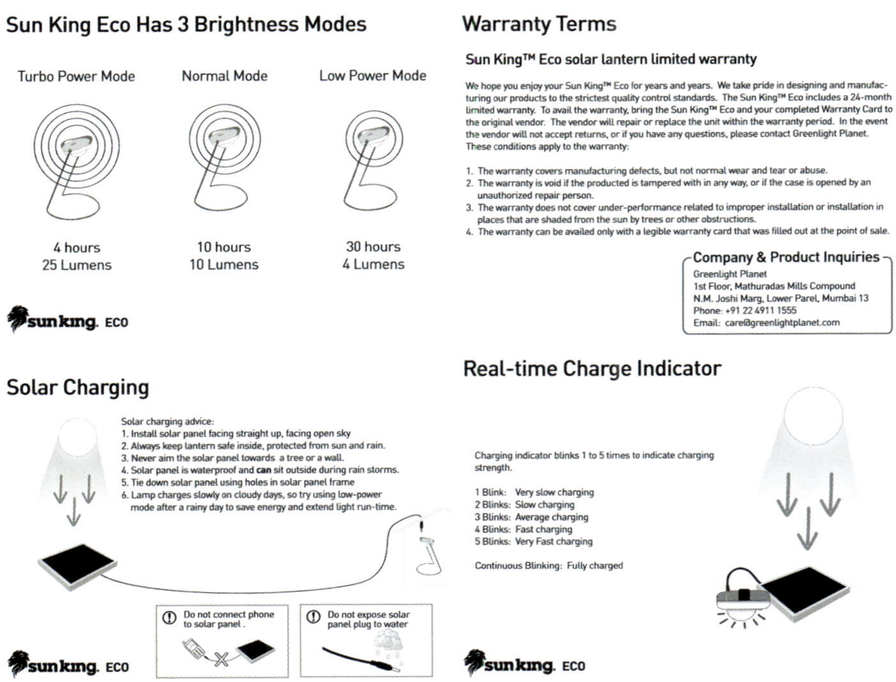

Figure 8.1
Instructions for how to use and recharge a pico-solar light. This example provides a good use of images to help explain to customers how to use and protect the product.

Source: Greenlight Planet

Figure 8.2 Keep modules free of dust and anything which prevents sunlight reaching the surface to maximise electricity generation. In this case, leaves and dirt reduce the amount of power being generated.

Source: John Keane

Figure 8.3 Customers need to know how to use systems properly to get the best out of them. For example, they need to know that modules need to be kept clean and that this can be done simply by using a damp cloth (no detergents).

Source: John Keane

Figure 8.4 Good products are built to withstand use in harsh conditions. No product is indestructible, however. This module shattered after falling face-down off a roof from a 5 metre height onto a hard surface.

Source: Gabriel Grimsditch

Understanding Benefits

It is important to be able to clearly articulate what benefits pico-solar systems offer customers. This means being able to explain:

- **The functionality of the product:** If it is designed to provide light and phone charging but will not charge larger products, make sure the customer understands this. Avoid any temptation to make claims about the system's ability to charge things which are beyond its power capability. This will only result in a disappointed customer.
- **The savings a customer can make:** Pico-solar products offer alternative energy options to traditional fuels such as kerosene, candles and disposable batteries and can save customers significant amounts of money over time. Helping customers calculate the savings they can make will increase demand.
- **Health and safety:** Pico-solar systems offer cleaner, safer and healthier energy options than flame-based fuels.
- **Convenience:** Pico-solar systems offer convenient access to energy and can power appliances at the flick of a switch. This can save people a lot of time and money, especially if they previously had to travel to the closest market in order to charge their phone.

Customer Service

It is extremely important for customers to have a positive experience of using pico-solar products so as to encourage further sales. Providing customers with information, customer support and strong levels of after-sales service will help ensure this. This means being able to answer customer questions, identify and honour any warranty issues and ensure customers know where they can purchase spare parts and have products repaired as necessary.

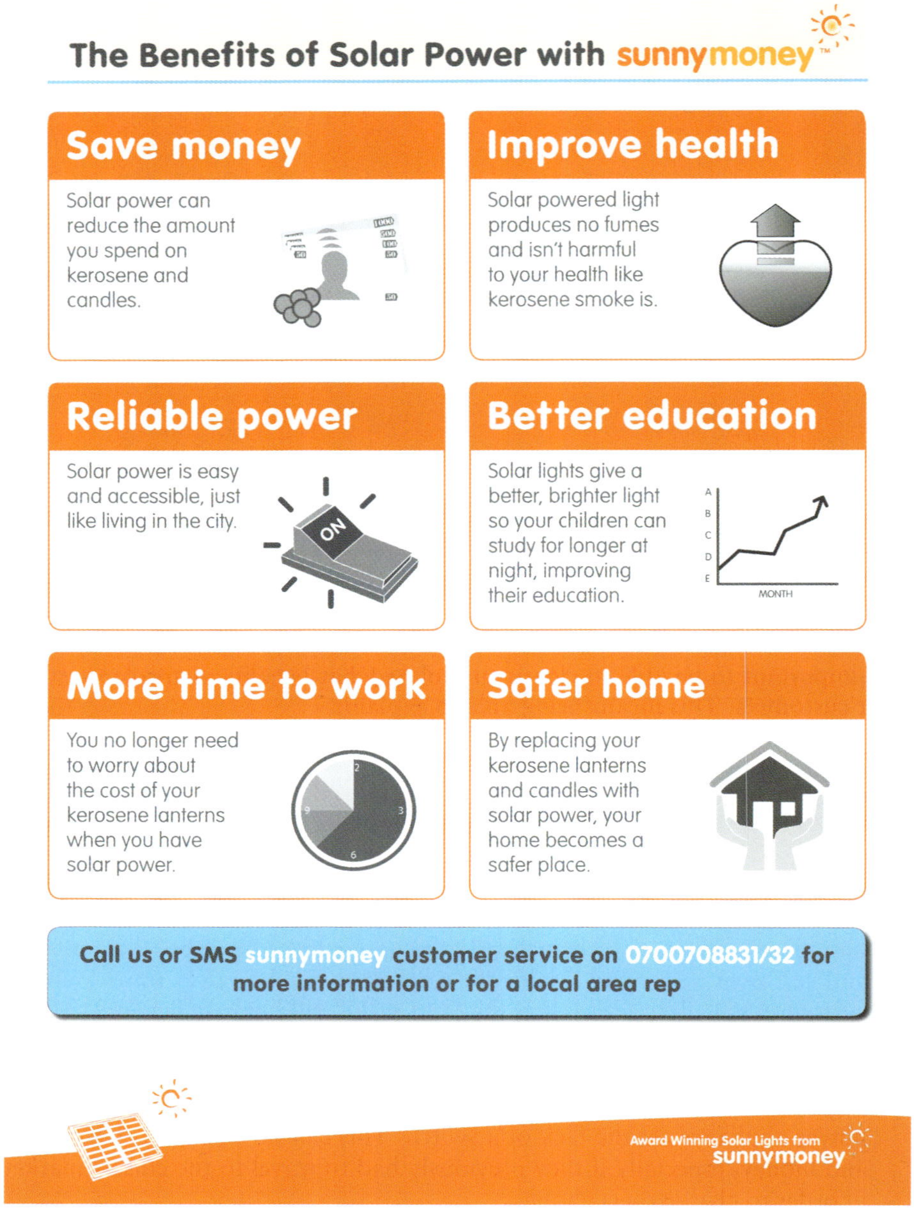

Figure 8.5 An advertisement outlining the benefits of using a pico-solar lantern instead of more traditional lighting sources such as kerosene.

Source: SunnyMoney

Warranties and Guarantees

If the product is returned by the customer within the warranty period, it is important to be able to identify if the issue with the product is due to misuse or a product defect covered by warranty. Retailers should ask any supplier what support they offer regarding warranties and get any agreement in writing when ordering products. A good agreement will mean that the supplier will replace any products with a warranty issue free of charge. If a product is returned by a customer:

- the first thing to do is ask the customer how they have been using and recharging the product so as to understand if they have been charging it properly and/or expecting it to provide more power than it is designed for;
- the product should then be placed in the sun for a full day of recharging to establish whether the issue is that it is simply out of power;
- if there is still a problem, refer to the manufacturer guidelines on any other simple inspections which can be conducted in order to ascertain the problem (additional guidance on troubleshooting is provided later in this chapter).

Spares and Repairs

Customers always want to know what they need to do in the event of a product needing repair or how they can access spare parts as necessary. A retailer therefore either needs to know where a product can be repaired or how to carry out repairs themselves. In order to carry out repairs, the repairman or woman must:

- be properly trained and have access to any manufacture product repair instructions and troubleshooting guidelines;
- have access to the correct spares to ensure the product is returned to full functionality.

Responsible suppliers will provide distributors with product trainings, materials and troubleshooting guidance. The battery is often the weakest part of any system. However, as battery types vary from product to product and many chemistries are not interchangeable, it is important to ask suppliers for access to spare batteries and to understand the costs.

The next section provides some general troubleshooting guidance to help assess what the problem is with a non-functioning product and what the solution may be. It also provides some examples of how to replace batteries and repair damage to faulty solar module cables and connections. As all products are different, however, always refer to the product manufacturer instructions for relevant instruction.

Troubleshooting

There are a number of things which can cause a product to stop working. It may be, for example, that the system is simply out of power and in need of a recharge. More serious issues can arise, however, such as faulty wiring, a broken module or an old battery no longer capable of storing energy. Table 8.2 provides a guide on how the source of the problem can be identified and the potential solutions available. This is intended as a guide only. Product manufactures should provide distributors with product specific trouble shooting guidance.

Table 8.2 Troubleshooting – potential problems and solutions

Problem	Possible cause	Solution
Product will not switch on or state of charge is low	Out of power – due to over use or lack of sunlight.	Recharge for at least one day in full sunlight. Many products have a SoC indicator to inform users of power levels. It is also possible to measure the battery voltage using a multimeter to assess SoC (see Figure 4.7, Chapter 4). A full day of charging should return the battery to a full SoC.
	Battery not accepting charge – this may be due to ageing battery or degradation	Replace battery
	Defective circuitry	Most pico-solar products use printed circuit boards, such that whole board will need replacement.
	Blown fuse	Replace fuse (if there is one to replace easily)
	Solar module not charging battery – this can be due to dirt covering or broken module, faulty wires/short circuits or lose connections (see Figure 8.7 for measuring module output)	Clean module if dirty. Repair module and cable or replace (see Figure 8.8)
	Switch defective	Ask technician to replace/repair switch
There is power in the system, but light will not turn on	CFL blown LED connection loose	Replace CFL Check for loose connections and repair
	LED blown (more common for LEDs to degrade slowly than blow)	Replace LED if possible
	Broken switch	Replace switch
	Faulty wire	Replace/repair wire

Tools needed: screwdriver, multimeter and a soft cloth.

Basic Repairs

Battery Replacement

The weakest part of any system is usually the battery, which may need to be replaced after a couple of years, depending on the battery chemistry. The signs of a dead battery are usually that it seems to recharge as normal (reaching the rated nominal voltage) but after a short period the voltage drops.

As batteries age, they also generally hold less charge than when new and it may be time to replace the battery.

While every system is different, it is fairly easy to open up most systems in order to access and replace the battery. Product specific instructions from the manufacturer must be followed to ensure the correct procedures are followed. An example of how to replace a battery is provided in Figure 8.6. The greatest challenge may be obtaining the correct replacement battery, rather than the physical act of replacing the battery. Retailers and distributors should plan ahead and ask manufactures for access to spare batteries to ensure the correct replacement battery is available when needed.

Example of Battery Replacement

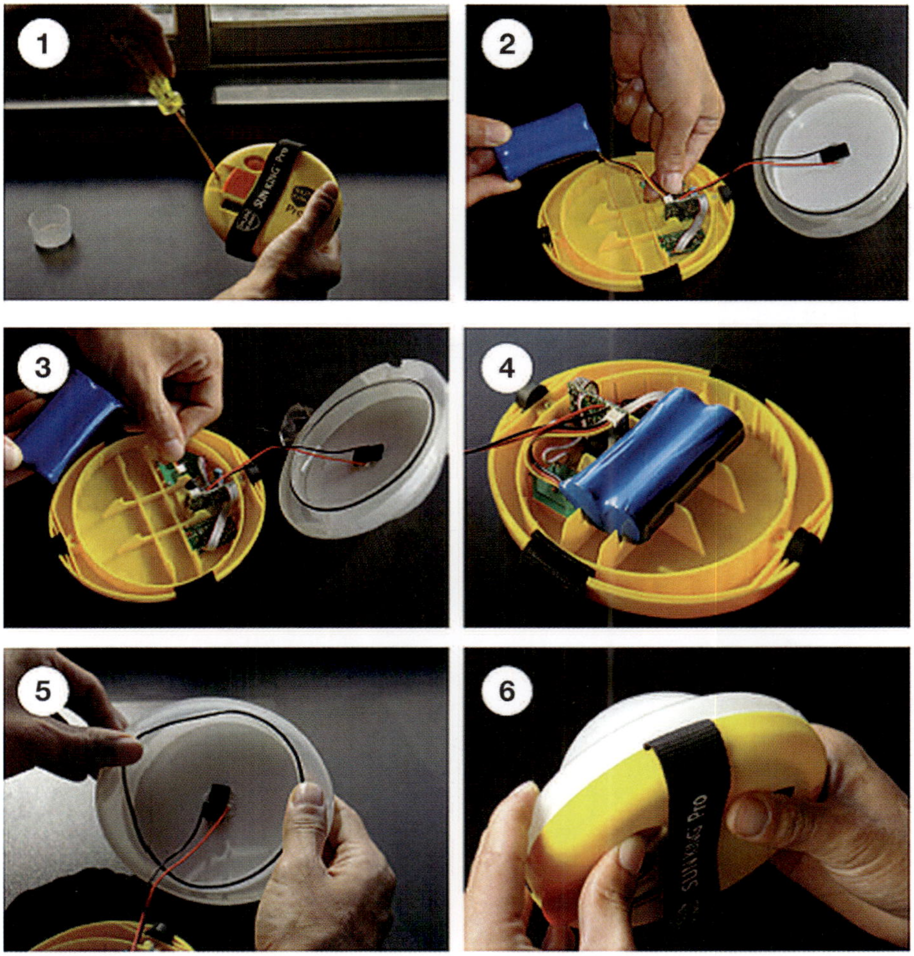

Figure 8.6 Replacing the battery in a Sun King Pro pico-solar light and phone charger

Tools required: small Phillips screwdriver

Time: 5 minutes

Step 1: Remove the two screws on the top of the lamp

Step 2: Carefully disconnect the battery connection from the circuit board

Step 3: Connect the new battery to the circuit board

Step 4: Ensure the new battery is properly positioned within the brackets

Step 5: Ensure the rubber O-ring is placed around the inside ridge of the lamp body and screw the lamp back together

Step 6: Test the light functionality. If the light does not turn on, charge it for 30 minutes in the sun and test again.

Source: Greenlight Planet

Once the battery has been replaced, it is important to dispose of the battery carefully so as to avoid any potential contamination to the surrounding environment. How a battery can be safely disposed of in a given geography is something any distributor, government or NGO should consider carefully when products are being used in remote areas with limited infrastructure.

Module Replacement or Repair

If a module is suspected of not working, place it in the sun and use a multimeter to test the Isc and Voc (see Chapter 3). If either of these shows a reading significantly below the rating of the module, there may be a problem with the module, the wiring or the connection between the two.

To test whether the problem is with the module or the wiring, carry out the same test by positioning the multimeter on the terminals at the back of the

Figure 8.7 Use a multimeter to test the Isc and Voc readings of the module by placing the 'probes' against the module's positive and negative terminals, i.e. carry out measurements without the module cables. Remember to position the module in the sunlight when measuring. If the module displays zero or below expected output, there may be a problem with it.

Source: John Keane

module. If the readings here show there is power, it is likely that there is an issue with the cable which can be easily repaired.

Replacing the Cable of a Pico-Solar Module

Replacing a Module Cable

Tools required:

- Flat head screwdriver
- Soldering iron with solder

Time: 20 minutes

Step 1: Remove the solar panel junction box cover by pressing down and sliding it outward. If the cover does not slide out under moderate pressure, use a flat head screwdriver to pry the cover off.

Step 2: Using the heated soldering iron, melt the solder binding the wire ends to the module solder points. Solder the replacement wire ends to the module contacts using fresh solder. Note:

- Soldering involves extreme heat which can damage the solar module and cause burns. It is important that the person soldering has experience of using soldering irons.
- Some modules have screw connectors which simplify the process of connecting and disconnecting cables with modules.

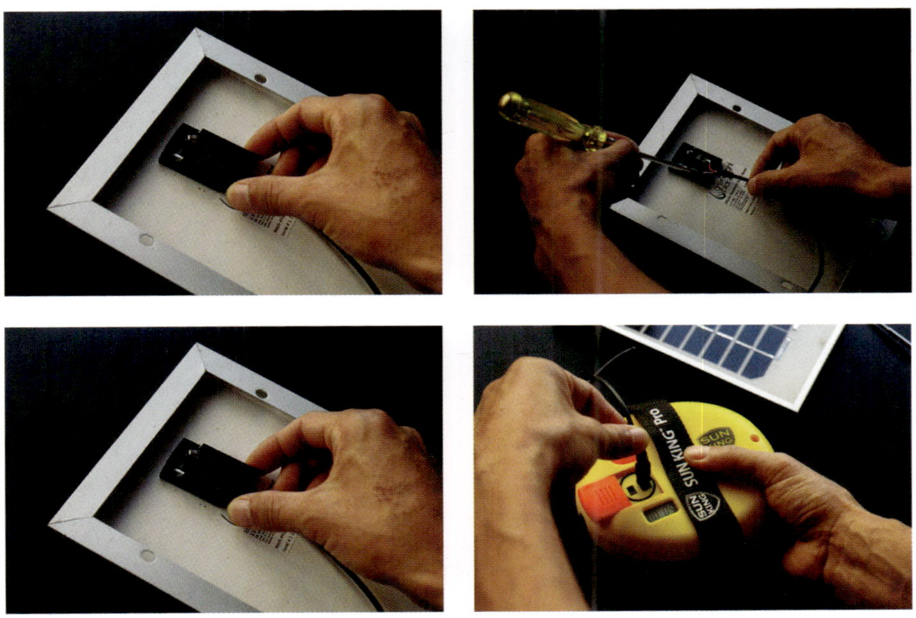

Figure 8.8 Replacing a module cable

Source: Greenlight Planet

Step 3: Replace the junction box by sliding it into place.

Step 4: To test, place the module in the sun and measure with a multimeter – or plug it into the lamp and verify that the unit is charging.

If the readings indicate that there is an issue with the module itself and it should be inspected for any obvious physical defect or breakage, it may still be possible to repair the module, but it is advisable to consult a technician at this point. If a module is beyond repair, it is possible to replace the module for systems which use a separate module. In this case, it is advisable to contact the manufacturer or, if this is not possible, source a solar module which has the same power ratings, specifications and jack as the broken module.

9 The Impact of Pico-Solar in Developing Countries

This chapter provides an overview of the positive and often transformative impact pico-solar systems can have on the lives of people living in parts of the world without access to electricity. With a particular focus on lighting, it begins by outlining the challenges faced by people who live without electricity and are often forced to rely on fuels such as kerosene for basic lighting. It then explains how the electricity pico-solar systems supply can save people money and have a positive impact on almost every aspect of life, in particular, health, education, the environment and the economy.

Light for Development

The importance of lighting on everyday life and the contribution that access to quality electric lighting can have on development is often substantially understated. It is clear, however, that access to clean, safe lighting is linked to economic and social development in multiple ways. A lack of reliable, quality lighting means that people are less able to carry out activities after dark, from conducting business and income generation activities, to reading and studying, to carrying out household chores and socialising. Lighting also plays an important security role. All of this is a particular problem across much of equatorial Asia and Africa where the sun sets between 6pm and 7pm throughout the year, leaving people in darkness for up to twelve hours each day.

The slow growth of electrification across many parts of the world, in particular Africa and Asia, is increasingly separating those with reliable lighting from those who lack it, leaving a substantial proportion of the world's population behind. It is therefore evident that access to energy and light is vital in order to reduce inequality and poverty. Recognising this, there is a growing movement that believes that access to electric light is a human right and that pico-solar systems can play a vital role in providing useful amounts of light and electricity to households which can have immediate impact on peoples' lives.

The Situation

There are over 1.6 billion people around the world with no access to electricity. In India alone, there are 289 million with no access to electricity and in sub-Saharan Africa, there is a rural electrification rate of just 14 per cent. A further one billion people live with unreliable or intermittent access to electricity.

This means that billions of people have no choice but to find alternative sources of energy for lighting. In much of Africa, kerosene lanterns are used for lighting, with candles and battery torches also used. Kerosene is also a common source of lighting in India. Families also burn firewood and biomass for light. Figure 9.1 shows a family in Zambia burning maize husks for some evening light. Figure 9.2 shows how people living in rural parts of northern Argentina mix

Figure 9.1 This family in rural Zambia is burning maize husks as a source of light. Maize is a staple crop across much of Africa.

Source: © Steve Woodward

Figure 9.2 Villagers in northern Argentina demonstrate how they often burn donkey manure mixed with animal fat as a light source in areas without electricity access or during power cuts. While the majority of Argentina's population enjoy access to electricity, there are still many communities with limited access in more remote areas.

Source: John Keane

donkey manure and animal fat for some evening light. Figure 9.3 is of a household using a kerosene lamp.

Due to the often unreliable nature of electricity access in developing countries, many households which are connected to electricity also resort to fuels like kerosene during power cuts.

The negative implications of the use of these sources of lighting are conversely linked to the benefits of using pico-solar products. These benefits are discussed in the next section.

Figure 9.3 Kerosene lights, which are toxic and dangerous, are used as the main lighting source by millions of households across Asia and Africa.

Source: © Steve Woodward

Impact of Pico-Solar

Pico-solar is a relatively new energy solution with only a small proportion of households owning a system in both Asia and Africa. While ownership is on the increase, in 2010, Lighting Africa estimated that 0.5 per cent of households in Africa owned a system, with Lighting Asia estimating a market penetration rate of only 4.5 per cent for pico-solar lanterns and SHS in India.

There are, however, many passionate advocates who have seen the difference pico-solar lights can have on lives. Raymond Serios, Energy Programme Director at the Negros Women for Tomorrow Foundation (NWFT) in the Philippines explains:

> Pico-solar lights address energy issues in rural communities, but the greater good that it does to families is that it liberates them from the limits they faced when they didn't have electricity. Imagine shorter study hours, less time for working on their business and less time spent talking with family; lack of evening light not only puts a cap on the family income but even negatively affects relationships. The effect of a pico-solar light is hard to quantify, but it definitely goes a long way — more than the economic or health benefits . . . it empowers people.

General well-being and some of the softer, but incredibly important, impacts of development programmes are hard to quantify. There is, nevertheless, a growing base of evidence that demonstrates how the use of pico-solar is contributing to social development and poverty alleviation. The areas of impact fit into five broad themes:

1. Economic
2. Health and safety
3. Education
4. Environment
5. Well-being

Economic

Household Savings

One of the most common benefits of pico-solar cited by users is how much money they can save through reductions on expenditure on kerosene for lighting. Research by the charity SolarAid indicates that in parts of Africa, spending on kerosene for lighting accounts for 10–15 per cent of household income. A pico-solar light can, therefore, help households escape poverty and spend money saved on basic needs such as nutritional food and sending children to school. In a small study conducted across Kenya, Malawi and Tanzania, SolarAid found that families who purchased a pico-solar light saved an average of around USD 70 a year through the resulting reduction in kerosene use. Over half of families included in this study stopped using kerosene lights completely after buying a pico-solar light. Entry level pico-solar lights retail for as little as USD 10, which means that in many cases, the product pays for itself within three months of use through the savings made on kerosene purchases.

Figure 9.4 This solar entrepreneur in Zambia shows off his shop sign which clearly explains the benefits of making the switch to solar lighting.

Source: © Steve Woodward

The SolarAid study also found that savings from reduced kerosene use were reported to be used on food for nearly half of respondents, many specifically mentioning being able to have a more balanced or nutritional diet. Investing in farming or business, and paying for school fees/costs were also common uses of savings, showing that increased savings opens opportunities for investment in the future for families.

Microsoft chairman and co-founder of the Bill & Melinda Gates Foundation, Bill Gates, believes that the cost of energy is inhibiting poverty alleviation activities: 'If you could pick just one thing to lower the price of, to reduce poverty, by far you would pick energy'.

Business Opportunities

As well as increased household income from reduced kerosene use, pico-solar lights also provide opportunities for small businesses to increase opening hours. A United Nations development programme study demonstrates that household businesses with improved lighting see up to a 30% increase in income due to increased productivity at night, with some pico-solar lighting companies reporting up to 50% increases in incomes due to the extended workday.

Health and Safety

Pico-solar can have a significant impact on health in multiple ways.

Indoor Air Pollution

Kerosene lanterns commonly cause fires and burns, result in poisoning, and also contribute to indoor air pollution. While there is plenty of evidence of the impact

Figure 9.5 Stella Mbewe using pico-solar to light her shop in Mafuta village, Chipata, Zambia. Pico-solar lights are being used by small businesses to attract more customers and enable shops to remain open after dark.

Source: © Steve Woodward

of indoor air pollution on health, there is limited data available for the use of kerosene specifically, a surprising fact when an estimated 290 million people use kerosene as their primary source of lighting in Africa and 380 million in India.

Lam *et al.* (2012) note in their review of household kerosene use:

> Well-documented kerosene hazards are poisonings, fires, and explosions. Less investigated are exposures to and risks from kerosene's combustion products. Some kerosene-using devices emit substantial amounts of fine particulates, carbon monoxide (CO), nitric oxides (NO_x), and sulphur dioxide (SO_2). Studies of kerosene used for cooking or lighting provide some evidence that emissions may impair lung function and increase infectious illness (including tuberculosis), asthma, and cancer risks.

There is a varying degree of awareness among populations of the health impact of using kerosene for lighting across Africa – some populations seem very aware of the issues whilst others either don't experience health issues or, more likely, do not trace their causes back to kerosene use. Of course, for many of the illnesses associated with kerosene use, there are numerous other factors that can also contribute.

Paul Shirima, a SolarAid pico-solar light customer in Kilimanjaro, Tanzania, says, 'We used to cough and get flu when we were using the kerosene lamp, also my children were getting eye pains because of the kerosene. We don't experience that anymore with use of solar.'

The results of poor health can also be economic. The World Bank (2008) quantified the potential economic loss from sickness from using kerosene lamps for light. As they explain, kerosene lamps emit particles that cause air pollution; burning a lamp can result in concentrations several times above the World Health Organization standard on air particles per hour.[1] The extra risk of respiratory sickness from exposure to these levels of particulate matter has been seen to result in an average of three lost adult work days each year, and it increases the mortality rate for children under five exposed to these fumes by 2.2 per 1000.

Health and Younger Children

It is often recognised that younger children suffer most from the negative health consequences of exposure to kerosene. A qualitative study conducted by the William Davidson Institute (WDI) at the University of Michigan which studied children eight and under, found that 'the youngest children (ages 0–5) suffered more headaches and reported more problems with eye irritation than children ages 6–8. This may be due to the time children in this age group spend sleeping near a kerosene lamp.'

Fires and Burns

Many of the alternative sources of lighting from electricity, such as kerosene lanterns, wood burning and candles, present a serious fire or burn risk. Sadly, there are many cases of injury and death caused by these methods, with regular reports from Asia and Africa of people burning to death after a kerosene lamp fell over. A particularly horrific example occurred in Tanzania in 2009 when a fire caused by a candle killed 12 girls at a boarding school.

Poisoning

Death or illness from kerosene can also be caused through poisoning, as Lam *et al.* (2012) note, 'Poisonings from ingestion of kerosene, particularly in children, are unfortunately common in developing countries. The problem is exacerbated by the common practice of insecure storage of small amounts of kerosene in soft-drink bottles without safety closures, often because purchasers of kerosene can only afford to buy a small amount at a time and provide their own containers for suppliers to fill.'[2]

Figure 9.6 The graves of 12 school girls in Tanzania who died following a fire in their school dormitory caused by a candle. Horrific sights such as this are all too common. Fires, burns and accidents caused by kerosene and candles occur on a daily basis across the world.

Source: © SolarAid

Figure 9.7 Kerosene is often purchased in small, affordable amounts and stored in old drink or water bottles. This creates a real danger of young children mistaking kerosene for a drink. Ingesting kerosene can be fatal.

Source: © Steve Woodward

Health Improvement Linked to Income

When pico-solar systems replace kerosene lanterns, they can have positive health impacts on households as well as reduce the risk of fires, burns and poisoning. Improved health can also increase a household's disposable income level. The aforementioned WDI study found that in some instances, savings from reduced purchase of kerosene are used to purchase food. This implies that savings can be spent on nutrition, which is an essential component to the physical and cognitive development of young children. According to the World Bank's Early Childhood Development website: 'Under-nutrition in young children can interrupt behavioural and cognitive development, educability, reproductive health, and future work productivity.' Malnourished children are more susceptible to illness and risk of death. In addition to these risks, the developmental delays caused by under-nutrition affect children's cognitive outcomes and productive potential as adults.

A SolarAid study also found that over 70 per cent of people surveyed noticed a change in the health of household members as a result of using the pico-solar light and reducing kerosene use; 40 per cent of these mentioned a reduction in coughing and chest problems, with a quarter noticing reduced eye problems and irritation.

Education

Education is often cited as one of the key drivers of development across the developing world, yet many people across the world are unable to study after dark due to lack of light. Pico-solar lights have the potential to change this situation.

SolarAid has seen that children in households with pico-solar lights study more after dark for a number of reasons, including: there are no additional running costs which limit use as is the case when burning kerosene or a candle; children do not experience eye strain in the same way they do with flame-based lighting; pico-solar lights do not emit fumes and can be used for longer without experiencing coughing or chest problems; students who own pico-solar lights are often more motivated to study than those without.

Joyce Mtei, a student at Mbalangasheni School in the Kilimanjaro region of Tanzania, is one of five children in her family. SolarAid research established that Joyce's father, Sinsolesa, bought a pico-solar light 'so as my children can study at night and also to reduce kerosene spending'. Before the purchase of the light, Sinsolesa explains that they used to spend nearly 9 per cent of their household income on kerosene. With the pico-solar light they don't use kerosene at all now and 'the surplus money has helped me to pay tuition fees for my children'.

Due to the perceived educational benefits of owning a pico-solar light, education authorities across East Africa are collaborating with the charity SolarAid and its social enterprise SunnyMoney to bring lights to students. From a small study conducted in Malawi, SolarAid found that nearly all teachers interviewed had noticed a difference in students with pico-solar lights. Two-thirds talked about increased or improved student performance with many specifically talking about improved pass rates and more students making the grade to enter secondary school. SolarAid has also found that lights appear to contribute to increases in school attendance, motivation, concentration and the amount of time spent studying. In addition to being able to do more preparation for their school classes by studying in the evening after sunset, WDI conclude that children using

Figure 9.8 Two students in Zambia sharing a pico-solar light to study after dark.

Source: © Steve Woodward

pico-solar lights develop higher levels of self-confidence about the quality of their work. Children who are prepared for class raise their hand in class more often and develop higher levels of self-efficacy in their ability to communicate about course concepts with peers and instructors. Due to the ability to study longer hours, children often develop improved aspirations for their future. With improved grades, research indicates that they wish to continue their education and also dream of better jobs.

Environment

Lighting Africa estimate that the cumulative effect of the off-grid community using kerosene and other biofuels for lighting contributes heavily to global carbon emissions; 100–150 million tons of CO_2 emissions. According to guidelines prepared for the UK Department of Energy and Climate Change (DECC), one litre of kerosene use relates to 2.53 kg of CO_2 emission; about 40 hours of lighting. Other estimates suggest that a kerosene lamp emits one tonne of CO_2 emissions over five years.

There has been some investigation into calculating the embodied energy – the energy required to produce the pico-solar lights – to establish whether the CO_2 emissions saved from reduced kerosene use through switching to pico-solar is less than the energy taken to produce and ship the pico-solar lights. These studies suggest that more CO_2 is saved by switching from kerosene use to pico-solar lights than is used in the manufacture and transportation of pico-solar lights to market.

Well-Being

Well-being is notoriously hard to define and measure as it can be very subjective. In addition to the positive impact pico-solar lighting can have on health, education and household budget, lighting can also mean families are able to spend more time together socialising. In short, lighting can extend the day, giving people choices, to spend more time playing, reading, socialising, studying and working, instead of going to bed when the sun goes down. There is also often a general feeling of progress and development for people who own a pico-solar light in rural, off-grid, areas. Pico-solar user Mphatso Gondwe from Malawi explains, 'I am very happy because the living standard of our family is more like one in town.'

Communications

For a long time, the radio has been the key medium of mass communication which acts as a source of news, information, education, training and entertainment to communities across the world. According to a UNESCO Education for all global monitoring report published in 2012, at least 75 per cent of households in developing countries have access to a radio. It is therefore difficult to overstate the importance of the radio to communities, especially those located in rural, remote parts of the world, which other mediums are unable to reach. Recent years have, however, seen the mobile phone challenge the radio's leading position

Figure 9.9 One pico-solar light providing light for a family of six people in rural Zambia. Households like this one can reduce the amount of money they spend on candles and use the money for things like education – or even more pico-solar lights. In a study by WDI one pico-solar customer in Tanzania specifically said that she frequently spends about two to three hours at the kitchen table with her children each night, helping them with their homework. This type of time investment in children is thought to be critical to the development of human capital. For more information, refer to Jonathan Guryan (2008) *Parental Education and Parental Time with Children*.

Source: © Steve Woodward

for mass communication, with an explosion in use and ownership, particularly across Asia and Africa, the same UNESCO report estimating that phones now cover over 70 per cent of the world's population. It is well documented that access to mobile phones is helping to transform the lives of many living in developing countries. While estimates vary, it is clear that billions of people across the world have access to mobile phones and radios.

Both radios and mobile phones only need relatively small amounts of electricity to operate. In areas where there is no electricity, however, keeping these devices turned on is the key challenge which faces radio and mobile phone users. In the case of radios, people are often forced to purchase disposable batteries each week, which becomes quite costly and pollutes local environments as batteries reach the end of their life. Mobile phone users, meanwhile, are often forced to travel to the closest town to access electricity and must often pay local businesses each time they want to charge up their phone. This takes up valuable time and costs money.

Owing a pico-solar system, capable of running a radio and charging a phone, can therefore save people a significant amount of time and money. Pico-solar also enables people to keep their radios turned on for longer, such that they no longer have to limit hours of use due to limited energy access, and keep their phones turned on, which means:

- Rural communities become less isolated and can communicate with family/friends living in the town or different areas; before mobile phones and in the absence of reliable postal services, people in rural Africa often communicated with people in other areas by giving notes to bus drivers to deliver.
- Farmers are able to gain information on accurate selling prices for produce and reduce the risk of receiving lower than acceptable payment for goods – a 2012 World Bank report, 'Mobilizing the Agricultural Value Chain', described mobile networks as 'a unique and unparalleled opportunity to give rural smallholders access to information that could transform their livelihoods'.
- Health workers are able to access information over the phone to improve support, training and aid diagnosis.

Figure 9.10 A farmer in rural Zambia charges his phone using a pico-solar light and phone charging product, enabling him to stay connected.

Source: © Steve Woodward

- Young adults are able to be informed of local news and opportunities.
- People who have never had a bank account are now able to transfer and save money on their mobile phones through mobile banking services such as M-PESA in Kenya.

Summary

The majority of people living without access to electricity across the world rely on expensive and outdated forms of energy – in particular candles and kerosene for lighting and disposable batteries for radios. Many people also have to spend time and money to travel significant distances to purchase the batteries, candles and kerosene. When the light bulb was invented, Thomas Edison opined that only the rich would continue to burn candles – yet more people are reliant on candles and kerosene for lighting today than were alive at his time.

The situation is similar for the rapidly increasing number of people who own mobile phones, who are forced to travel to the closest town to access electricity, often paying businesses each time they want to charge their handset. This means that the poorest people on earth are spending significant sums of money – sometimes up to a third of their household budget, and time – often travelling for hours each week, on energy solutions. This is valuable money and time which could be put to better use if people had access to pico-solar light and charging solutions.

This chapter has outlined the positive impact of replacing the current energy solutions with pico-solar, from improved health, safety and education conditions to reductions in household energy expenditure, time savings and improved quality of life. Pico-solar systems and the small, but incredibly useful, amounts of electricity they generate, can clearly, therefore, play an important and integral

Figure 9.11 This image is from a poster produced by the Lighting Africa programme which shows how children without access to electricity struggle to study after dark and how different life can be once they have access to electric lighting.

Source: © Lighting Africa

role in tackling poverty at household level. While it is important to cite statistics which demonstrate the scale and degree of the problems facing people who live without access to electricity and the benefits access to a pico-solar system can bring, images (see Figure 9.11) are often the simplest, most effective, mediums to explain the situation.

As more households start to use pico-solar solutions, the benefits of access to small but very useful amounts of electricity, can be extrapolated across large populations and have significant, positive, socio-economic and environmental impacts on whole nations, regions and continents.

Notes

1 Kerosene lamps emit particles that cause air pollution; these are measured by the concentration of the smallest particles per cubic meter (PM10). Burning a litre of kerosene emits PM51 mg/hour, which is just above the World Health Organization 24-hour mean standard of PM10 of 50 mg/m^3. But these particles do not disperse, so burning a lamp for four hours can result in concentrations several times the World Health Organization standard. The extra risk of respiratory sickness from exposure to these levels of PM10 is captured in the hazard ratio (the relative probability of the exposed versus unexposed being sick), which is 3.5.

2 Nicholas L. Lam, Kirk R. Smith, Alison Gauthier and Michael N. Bates (2012): 'Kerosene: A Review of Household Uses and their Hazards in Low- and Middle-Income Countries', *Journal of Toxicology and Environmental Health*, Part B: Critical Reviews, 15:6, 396–432.

10

Selling Pico-Solar at the Base of the Pyramid

Market Challenges and Solutions

The largest potential market for pico-solar products is the world's off-grid population, people who live without access to electricity or those living in areas where electricity supplies are intermittent and unreliable. This market, which is predominantly composed of low income households and commonly referred to by economists and development actors as the Base of the Pyramid (BoP), is often difficult and expensive for companies to reach. This chapter examines the ways the sector is working to overcome challenges of:

- logistics and distribution;
- financial barriers which prevent customers from purchasing pico-solar products;
- creating demand for and trust in pico-solar companies and products;
- ensuring customers receive good service;
- taxes and tariffs;
- product disposal, reuse and recycling.

It summarises what NGOs and governments can do and should not do, to help facilitate the growth of this market. This chapter will be of particular interest to companies, governments and NGOs involved in promoting the sale and distribution of pico-solar products, as well as students who wish to learn more about the challenges in reaching these customers.

Challenges in Reaching the BoP Market

The BoP market is made up of low income households across the world, many of which do not have access to electricity and are located in rural areas with limited infrastructure. It is therefore, by definition, a market with its own set of particular needs and characteristics. A study conducted by the consultancy firm Hystra, which reviewed several organisations selling goods at the BoP, concluded that families do not usually have the cash at hand to invest in new products, are wary of unknown technology and of the unavailability and/or cost of maintenance services or replacement parts. The study also established that the downstream ecosystem in terms of distributors, financiers and maintenance services generally do not exist or are insufficient. Figure 10.1 provides an excellent summary of the many challenges facing the pico-solar industry at the BoP:

	Challenges	Successful business model characteristics
R&D/product design	• Competition for R&D resources traditionally focused on more developed markets • Lack of consumer might, needed to design products to match local needs and expectations • Developing products that are affordable but desirable for target consumers	1. Localize corporate R&D efforts in BoP markets 2. Develop aspirational rather than functional products 3. Acquire locally produced products or players 4. Partner for joint product development
Manufacture & supply chain	• Low availability and/or reliability of quality inputs • Lack of reliable local manufacturing partners • Limited access to growth capital to scale operations and realize cost savings	1. Import finished products 2. Import parts and assembled locally 3. Local production by local players either own or contractor of large companies 4. Development of sustainable fuel supply chain for stoves
Marketing Distribution	• Limited and/or fragmented marketing channels • Low consumer awareness/understanding of Ide-cycle costs and benefits of energy solutions • High cost to maintain physical presence needed to support local partners and build/maintain brand • Fragmented / diffuse distribution channels with limited technical capacity • Limited physical supply and distribution infrastructure • High cost of (working) capital for distribution partners	1. Distributor - Dealer network 2. Institutional Partnership 3. Franchise 4. Rental / Leasing system 5. Own distribution / Direct to consumer
Customer finance	• Limited availability or awareness of financial services for BoP customers • High financing costs due to (perceived) high risks of lending to BoP customers	1. Partnering with microfinance institutions 2. Customized last-mile billing system 3. Leverage channel-sharing system to collect payment 4. Leverage carbon finance to subsidize/finance customers
Maintenance & end of product life	• Limited availability of after-sales support partners with necessary technical or commercial skills • Undeveloped supply chain for replacement parts • Limited or no waste management infrastructure	1. Done by large companies directly 2. Subcontracted or outsourced to local players 3. Train local people to do it and support the development of small businesses 4. Use environmentally friendly materials and/or build product/ material recycling into business model

Figure 10.1 Challenges facing the pico-solar industry. In order to be successful, the pico-solar industry needs: (1) well designed, quality, products; (2) robust supply chains; (3) profitable distribution models; (4) effective marketing strategies to increase consumer trust and awareness; (5) ways to overcome the finance barrier which can prevent customers from purchasing products and finally (6) systems which ensure customers receive reliable after-sales support.

Source: Business Solutions to enable energy access for all. The WBCSD Access to Energy Initiative. WBCSD Development

Logistics and Supply Challenges

There are many challenges to consider when supplying pico-solar systems markets across the world. Significant ones include: the time it can take to ship products internationally, deal with customs and also set up distribution networks for and transport goods to rural markets where many potential customers live. Getting the timing of shipments right is not always as simple as it may sound. The majority of pico-solar products are made in Asia, specifically in India and China. In some cases, it can take up to six months for products to travel internationally from the factory to the target customer in a rural area on a different continent. Time is important for a number of reasons:

1. Many potential customers have seasonal incomes based on local agricultural economies, which means that they may only have sufficient funds available to purchase a pico-solar system at certain times in the year.
2. Some battery chemistries degrade significantly over a period of months such that retailers should need to check that batteries are in a full SoC before products are sold and recharge them before sale if this is not the case.

The first issue underlines the importance of distributors getting the timing right for any imports, so that they can market products to rural customers when money is available in the local economy. Missing this window, due to delays experienced in shipping or at customs, for example, could lead to a distributor being in the unenviable position of having a large volume of unsold stock that must be stored until the next harvest has completed – which could be months away. Not only

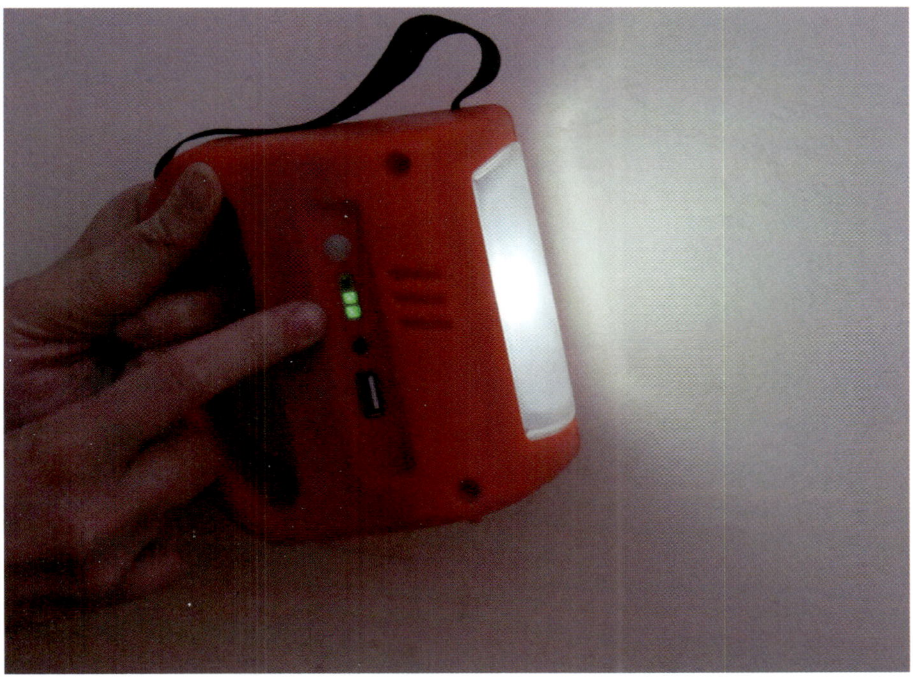

Figure 10.2 The SoC indicator on this product shows that the product is not fully charged. Retailers and customers alike should ensure that the product is fully charged before it is sold.

Source: © Kat Harrison

does this tie up a lot of cash in stock, with the possible result of cash flow pressures, it also increases the possibility of stock degrading over time.

It is also worth noting here that production in China effectively closes down for a month each year around the time of the Chinese New Year which takes place between 21 January and 20 February. It is typically not possible to organise shipments of products in large volumes over New Year. In short, there are many variables to consider when moving large volumes of pico-solar products across the globe. Figure 10.3 provides a theoretical example of how long it can take for a product to travel from a factory in China to the point of sale in Africa.

One solution which can ease the situation is to send products by air, as opposed to sea. While this tends to be much quicker, it can add significant costs, leading either to higher prices, making it more difficult for low income households to afford the products, or lower profit margins for the business supplying the products – making the enterprise less viable. The flow chart below provides a simplified example of how long it can take for a pico-solar product to reach the market from the day of order in China to the point of sale in East Africa. In this example, it takes over four months. In reality, times can and do vary significantly. The method of distribution and retail chosen as well as the number of middlemen also impacts on timing.

Figure 10.3 From factory (China) to retail outlet (East Africa): an example of how long the journey can take.

Source: John Keane

1. Manufacturing process can be quicker or longer, depending on the size of order and from manufacturer to manufacturer.
2. In theory, pre-shipment verfication should not take long, provided all documentation is correct. Incomplete documentation can lead to delays.
3. International shipping by sea from China to Africa can take less time than this, but delays are not uncommon.
4. In theory, customs requirements should not take long if all documentation is correct and there are no disagreements on how the goods are classified by the revenue authority. Delays can occur due to a backlog of imports, however.
5. Internal transport times vary considerably depending on distribution methods, sales models and geography.

An example of how challenging international logistics can be is when ordering products from China to arrive in Malawi in time for the harvest season in April – when customers have more money available to spend on products. Under normal circumstances, the Malawian distributor will need to place an order at least three months in advance to ensure products arrive in time. In this case, however, the distributor needs to also take into account the Chinese New Year, which takes place in January or February. At Chinese New Year, production lines grind to a halt for weeks and any orders not received well in advance of the New Year break may be significantly delayed, arriving in Malawi after the main selling period. Careful planning is therefore needed to ensure products arrive on time.

Many logistical challenges involved with shipping goods internationally can be overcome if products are manufactured locally. India is a good example of a country with a significant off-grid market and its own pico-solar manufacturing capacity. Manufacture in Africa, one of the world's largest markets, is limited, with only a small number of examples of product assembly that relies on key components being imported.

Distribution

Once products are in the country of sale, the next challenge is distribution. Distribution may be defined as the manner in which goods move from the manufacturer to the point where the consumer can buy them. The traditional distribution model has three main levels: the manufacturer, the wholesaler and the retailer. The challenge for distributing pico-solar systems, however, is that the target market is often located in remote rural, unelectrified, areas of the world.

Figure 10.4 A pico-solar assembly line in Mozambique. The majority of pico-solar products are made in Asia. There are a small number of examples where products are assembled in Africa. The advantages of assembling locally include potential tax incentives for companies and the ability for products to be repaired, and possibly partially recycled, more easily. Key challenges include limited manufacturing infrastructures, which means that most components need to be imported. Also, transportation of goods across borders in Africa can be more costly than importing direct from Asia due to limited infrastructure and import/export taxes. Chapter 11 includes a case study (p. 151) that provides an example of assembly units being set up in Mozambique.

Source: fosera

This section discusses some of the key challenges the industry faces in distributing pico-solar systems and highlights channels which are currently being used to reach the market.

The Pico-Solar Market

A significant proportion of the pico-solar market is located in rural, not easily accessible and therefore expensive to reach, parts of the world. It is also a market dominated by low income households which, by definition, have less money available to spend on consumer products. Therefore, in order for the market to reach its true potential, companies need to establish efficient and effective models of distribution that enable products to be sold at an affordable price – while still making a profit for each part of the value chain. The importance of overcoming the challenge of distribution is highlighted by Coimbatore K. Prahalad, author of the renowned book, *The Fortune at the Bottom of the Pyramid* (2006): 'Innovations in distribution are as critical as products and process innovations and designing methods for accessing the poor at low cost is critical.'

A study carried out by Dalberg Global Development Advisors in 2010, concluded, 'We have seen no "magic" solutions to the problem of distributing solar portable lights at scale in Africa.' The study does go on to suggest a number of ideas for reducing distribution cost and leap frogging 'last mile' distribution challenges, which include targeting worker's cooperatives and mobile phone operator networks. Developing viable, scalable methods of distribution is clearly one of the greatest challenges currently facing the pico-solar industry. Fortunately, a number of organisations have been working hard to demonstrate that it is possible to sell pico-solar products in large volumes and reach people living in rural areas. Some leading examples are discussed as case studies in Chapter 12. Figure 10.5 outlines five key distribution strategies which can be used to reach customers.

1. Proprietary distribution: This is when the manufacturer develops and owns its own dedicated retail distribution networks. This is expensive to set up, manage and control, and few companies try it.

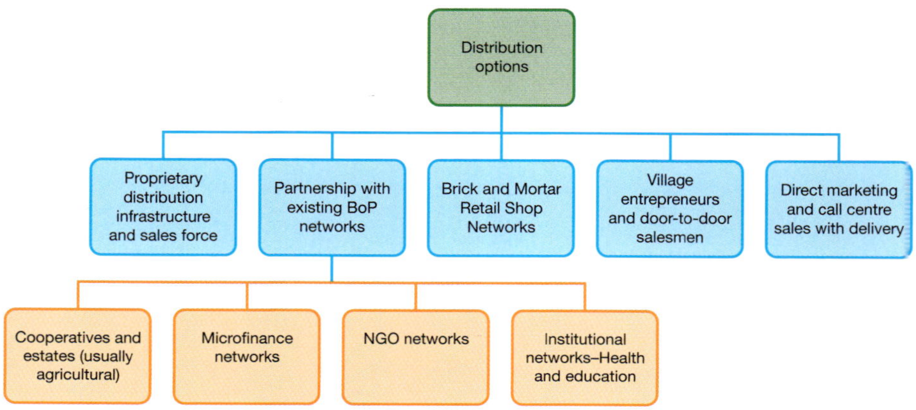

Figure 10.5 Distribution options for pico-solar products at the BoP

Source: John Keane

2. Partnerships with existing networks: There are many existing networks which work with and serve rural populations which can offer the potential of easy access to customers.
3. Retail outlets: Brick and mortar shops exist independently and as networks throughout the world and can incorporate pico-solar systems as an additional product line.
4. Village entrepreneurs: Door-to-door salesmen and women. These can be independent entrepreneurs or supported to work as franchisees who sell within local communities.
5. Direct sales delivery/catalogue order: In response to direct marketing, such as marketing done from a call centre, this can involve customers placing and paying for orders remotely – for example, through mobile phone banking. Products may then be delivered via local buses on receipt of payment.

There are many advantages, weaknesses and challenges and opportunities involved with each of the distribution strategies in Figure 10.5. These are summarised in Table 10.1.

Need for Efficient Value Chain

Achieving profitability, while at the same time keeping products at a price that is affordable for the low income households, is a key challenge facing the sector. Careful attention must therefore be placed on ensuring that any distribution system is efficient, while at the same time affording each part of the value chain sufficient margin to be profitable. Figure 10.6 shows a typical value chain.

As we will see in the next section on financial barriers and upfront costs that prevent many customers from purchasing a solar product, there are a number of potential solutions to this problem, such as 'Pay As You Go' (PAYG) products. PAYG systems enable products to be rented or purchased in installments, thereby reducing financial barriers to product uptake, while at the same time offering the potential to increase overall profit margins.

Financial Barriers and Solutions

There are two principal financial barriers facing the pico-solar sector which need to be overcome. First, many companies operating in the sector need more investment and working capital in order to scale up. Second, the upfront costs involved in purchasing a pico-solar product prevent many potential customers from being able to afford them.

Financial Constraints – Manufacturers, Distributors and Retailers

With a number of exceptions, the pico-solar industry is characterised by relatively small manufacturers, distributors and retailers with fairly limited access to finance, or by larger companies not yet investing heavily in a new market which may be seen as fairly expensive to reach, risky and not as high a priority as more

Table 10.1 Distribution channels: strengths, weaknesses and opportunities

Channel	Strength	Weakness	Opportunity
Proprietary distribution	Offers manufactures direct route to customers with potential for less middlemen	Can be expensive to set up and challenging to manage Only one brand of lights sold – reducing product options available	More profit potential for the company
Retail outlets/ dealers	Established networks of shops exist throughout world Easy to deliver stock to physical outlets May have access to capital to purchase stock	Not always close to market – shop owners rarely take proactive approach to reaching rural customers In absence of proactive approach, there is a risk of products gathering dust in shops as customers not aware they exist (unarticulated need)	Well placed to serve the market once it has momentum Opportunity to stock large networks – e.g. mobile phone outlets and petrol stations
Rural entrepreneurs	Close to the customer Understands market Can encourage word of mouth Can deliver after-sales service	Not always established and can be expensive to set up and train Managing, supporting and delivering products to a widespread, hard to reach, network of entrepreneurs can be difficult Low access to capital	Mobile phone makes communication easier Mobile money makes transactions easier Low cost motorbikes entering market offer entrepreneurs improved access to stock and customers
Existing networks e.g Agricultural cooperatives School networks	Established Easy access to members	Only serves a small proportion of the market Network not owned or controlled by pico-solar distributor and may not be set up for retail Works best if mission of network is aligned with the benefits a pico-solar product brings e.g. light for education	Opportunity to reach large volumes of people quickly

Channel	Strength	Weakness	Opportunity
Microfinance institutions (MFIs)	Established, easy access to members and clients Can finance customers, overcoming cost barriers Finance for entrepreneurs who want to start a solar business	Serves only 5–10% of market Low revenue products can be less interesting to financiers	Partnering with successful distribution models could unlock more customers for MFIs
Direct delivery	Opportunity to reach customers cheaply and efficiently through phone, text messages and leaflets	Requires access to customer database or network to reach them and may require middleman to facilitate sale Customers need to trust company to part with cash before receiving a product Needs local delivery point	Mobile money can facilitate remote payment As smartphones become more ubiquitous, opportunity to reach customers online and launch online delivery service

Whichever method of distribution is used to bring pico-solar products to the market, in order for it to succeed and be sustainable, there needs to be sufficient demand for the products at prices which make the operation profitable throughout the value chain.

Figure 10.6 Stages in the value chain as a pico-solar product travels to market.

Adapted from source: Solar Lighting for the Base of the Pyramid – Overview of an Emerging Market. Lighting Africa Report 2010

traditional markets in industrialised nations. The main finance needs facing manufacturers, distributors and retailers are summarised below:

Manufacturers: The majority of manufacturers operating in the pico-solar market are still relatively small and need to:

- attract finance to enable them to carry out ongoing research and development in a fast moving sector with technology improving and changing all the time;
- secure working capital in order to scale up production as product orders rapidly increase.

The main solutions for manufacturers include: improved trade finance terms with component suppliers, equity investment, bank loans and guarantees.

Distributors: The key financial challenges facing small distribution companies which specialise in selling to rural customers are:

- the development of innovative, 'last mile', distribution strategies that can be scaled up costs money, requiring grants or investment – which is often seen as risky;
- working capital – it can take many months for products to travel from the factory gates to the customer. This means that working capital is 'tied up' for many months at a time.

The principal solutions for distributors include: improved trade finance terms with suppliers, equity investment, bank loans and guarantees, philanthropic capital and donations which recognise social impact.

Retailers: Many retailers have limited access to working capital which means:

- they cannot purchase stock from distributors in bulk, adding to costs (smaller purchases generally mean higher per unit transport costs and less discounts from distributors);
- they cannot extend credit to customers without the involvement of a third party – such as a microfinance organisation. Note: Even if retailers are in a financial position to extend credit, it is often seen as too risky in less formal markets.

The solutions for retailers can include: improved trading terms with distributors – although this is risky and places additional cash constraints on distributors, microfinance bank loans and philanthropic capital.

Finance Barriers and Solutions for Customers

The upfront costs associated with purchasing large solar systems are a well-documented barrier that can prevent customers in western markets from opting for solar solutions. Government intervention and the creation of favourable market conditions through initiatives such as feed-in-tariffs (FiTs) have helped overcome these financial barriers and incentivised many people to install solar systems in their homes. Pico-solar products, designed for low income households across the world, while only a small fraction of the cost of their larger counterparts at prices of up to USD 100, often present similar financial barriers to their target customers, many of whom live on a dollar a day.

The good news for customers is that the price of solar modules has plummeted over recent years to less than USD 2/W in 2012. In the case of pico-solar, a 2012 study by Dalberg forecasts a continuing decline in pico-solar product prices, driven by falling prices of solar modules, batteries and LEDs. While declining prices are good news for customers, they do not solve the 'upfront cost' issue. This is an issue which can only be overcome through offering customers some form of finance solution.

Overcoming Upfront Costs

Across much of Africa, Asia and other low income parts of the world, the key barrier to being able to purchase a product or service is price. Across the world, companies are continually developing new ways of selling their products and services to as many customers as possible.

Companies have responded by offering their product or service in smaller quantities than they normally would in richer nations. This enables customers to purchase, for example, shampoo or alcohol in 50 ml sachets for 20 US cents, rather than a half litre bottle for a couple of dollars. Similarly, mobile phone companies sell talk time scratch cards for as little as a few cents, thereby enabling poorer customers to use their service.

In the world of lighting, people across the world often purchase a light outright, but then pay for the fuel in small portions – whether it is a kerosene light or small hand-held torch. Battery powered LED torches and lanterns, popular in countries such as Zambia, for example, enable customers to purchase a lantern for as little as a couple of dollars and then purchase cheap, low capacity batteries for around 20 cents each, weekly or as needed. Kerosene lights work in the same way: the customer buys the light, which may be a USD 5 mass-produced hurricane light or light made locally out of recycled food tins which retails for as little as 25 cents. The customer then purchases kerosene in small, affordable amounts, as needed.

While pico-solar systems can be far cheaper than a traditional SHS, they are not generally available on the market for anything less than USD 10 for the smallest entry level pico-solar light. This still represents relatively high upfront costs, preventing many low income households from purchasing a pico-solar system – and it is not really possible to cut an already small pico-solar system into smaller amounts to make it more affordable. Nevertheless, a wide number of solutions aimed at overcoming this financial barrier have been developed over recent years.

Figure 10.7 This photo is of a small shop in rural Zambia, now lit with a pico-solar system. Kerosene for sale can be seen at the bottom right of the photo. It is stored in a large yellow container. Customers often come to the shop with their own 0.5 litre plastic water bottle to purchase kerosene in small, affordable amounts.

Source: © Steve Woodward

Figure 10.8 Shopkeepers in rural Zambia with their new pico-solar light which enables them to trade after dark and means that they do not waste money on disposable torch batteries or kerosene. This photo shows how products such as washing soap are sold in small packets which are more affordable than the larger boxes seen in stores in wealthier parts of the world. This shop also sells toothbrushes, which are often poor quality – the author knows from experience that they can snap during their first use. It is a similar story with disposable batteries, which are often quite cheap and may only last for a day or so. These are examples of how people living on low incomes can only afford to purchase less expensive goods, but these are often of extremely poor quality so that people end up wasting money.

Source: © Steve Woodward

Pay as You Go (PAYG) Pico-Solar

The latest innovation which offers a solution to the finance problem comes in the form of PAYG pico-solar. This enables customers to purchase or rent a pico-solar system by making a small down payment, a fraction of the total system cost, and then periodically 'top-up' the system with a small payment. If the customer fails to 'top-up' the system as required, the supplier of the system can turn off the pico-solar unit remotely, courtesy of a small chip which is integrated into the system's circuitry. At the time of writing there are an increasing number of companies entering the PAYG pico-solar market with a range of competing products, technologies and business models.

In Kenya, the company M-KOPA has teamed up with a leading manufacturer to offer a three light pico-solar system to the customer for an upfront cost of around USD 30. The customer can then 'top-up' the system each day, week or month – payment terms are flexible – by sending money through the mobile money solution M-PESA, itself another revolutionary innovation which has transformed the way people handle money in the country. In this case, each 'top-up' goes towards purchasing the system over time so that after around two years, customers will have paid off the system. Of course, they will have paid more for the system than they would have paid if they had purchased the system outright at the beginning, but this is in line with traditional financing systems. Once the agreed payment amount has been reached, M-KOPA then unlock the system remotely and the customer is the proud owner of a multi-light pico-solar system. Should the customer stop topping-up the system before it has been paid off, M-KOPA is able to turn off the system remotely until the customer starts paying again, much like the electricity company can turn off a customer's supply if they don't pay their bills.

The challenge here is how to reclaim a small pico-solar system from a customer who stops paying or to ensure that a customer continues to pay as agreed. Companies like M-KOPA have been working hard to set up systems which reduce the likelihood of customers failing to make payments. In this case, as systems are topped up through the popular mobile money system, M-PESA, this reduces the likelihood of a customer disappearing without paying for a system as it may also mean that they can no longer use the mobile money service, which is useful to the customer for many other things, such as sending and receiving money.

Azuri, another PAYG company in Kenya, offers a similar pico-solar system to the customer, with similar payment terms. In Azuri's case, the customer can purchase scratch cards and input a code into a keypad on the pico-solar system to activate it for a set amount of time. Both of these companies are enabling customers to start using multi-light pico-solar systems for a small down payment of around USD 10 and using technology to allow the customer to continue paying for the system over time. Figure 10.9 and the following text shows a typical PAYG process:

Step 1: The initial down payment in Step 1 is typically only a small fraction of the total product cost, thereby helping to address upfront product cost barriers.

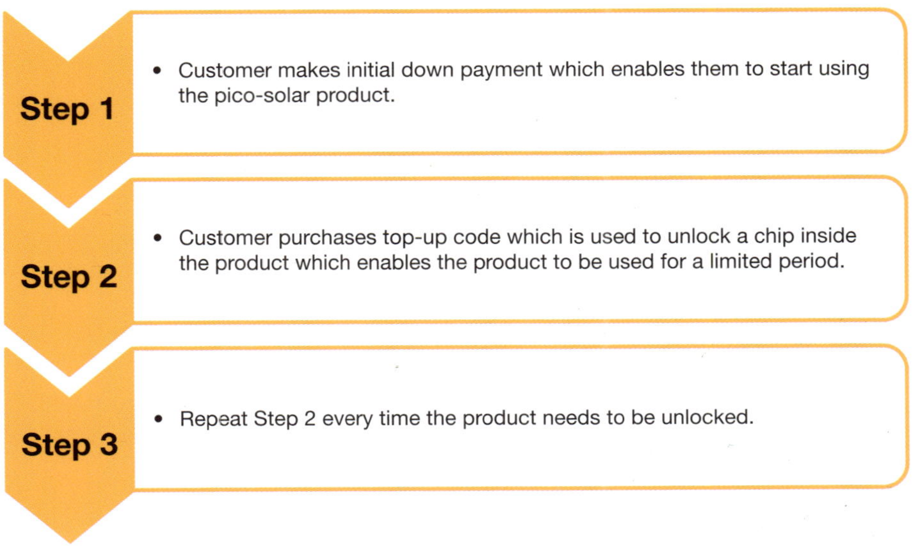

Figure 10.9 How a typical PAYG model works.

Source: John Keane

The actual deposit varies depending on the size of the product and the business model.

Step 2: Top-up payments can be small amounts, similar to what a customer may have spent on kerosene lighting or phone charging for the same period. Alternatively, the top-up payment can be one of a small number of larger payments which enable the customer to pay for the product in a few instalments. Top-up codes may be purchased through mobile money or via scratch cards.

Step 3: PAYG business models vary. Some companies offer customers the opportunity to purchase a product by making a small, limited number of instalments over a short period of a few months. Other companies have opted for longer-term repayment plans of up to two years and include options for customers to upgrade to larger products over time. Companies that adopt models which require customers to continue making payments over long periods need to ensure that the risks of customers stopping payments during that period are kept to a minimum. For example, it is important that the product remains operational over the two year period and customers remain satisfied with the service provided by the company to ensure that the customer continues making payments. PAYG can also be used to facilitate the rental of products.

All PAYG businesses are designed to reduce the upfront costs of purchasing or accessing a pico-solar product. This does mean, however, that the customer will ultimately pay more for the product over time than if they had bought it upfront in one go – but this must be viewed as the cost of financing.

There are obvious potential risks when offering credit in this way. While technology enables the company to turn off a system should a customer stop paying, it is potentially quite costly to try and reclaim a system from a family

Figure 10.10 Azuri keypad and Azuri scratch cards. This image shows one type of pico-solar PAYG system which can be topped-up by typing a code from a scratch card directly into a keypad on the product.

Source: John Keane

living in a remote village. There are also many variables which could reduce the ability of a customer to pay, such as a poor harvest and unexpected expenses. Any PAYG company therefore has to assess and factor in these risks to their business model and introduce measures to mitigate potential problems and account for potential pressures on cash flow. Conversely, as the customer is paying for the pico-solar system over time, in the examples given, the companies have an obligation to ensure that the system continues to function correctly during the lifetime of the agreement.

While the above examples feature multi-light systems purchased over extended periods of up to two years, which effectively require the company to act as a utility service, a number of companies are looking at introducing PAYG technology into smaller, entry level, pico-solar lights, which normally retail for between USD 10 and USD 40. The rationale is simple, if USD 10 is still too much for a potential customer to pay, it is possible to use PAYG technology to allow that customer to purchase the light in a number of small instalments, such as three USD 4 instalments, thereby further reducing the entry cost of owning a pico-solar light.

Rental Solutions

Offering products to customers on a rental basis as opposed to selling them is another way to overcome financial barriers. Renting systems can be administratively challenging and exposes the entity renting out systems to potential risk. Measures must be put in place to lessen the risk of systems not being returned.

A number of rental examples exist around the world, for example in Laos, where the company Sunlabob has set up projects that enable entrepreneurs to rent pico-solar lights to customers at prices similar to that of kerosene. While the entrepreneur manages the light rentals, the lights themselves are actually owned by the community, with a percentage of revenue from the rental system going into a maintenance fund which is then managed by a village energy committee. There are many challenges with setting up rental systems, such as:

- They typically require an entrepreneur to have access to relatively large amounts of money in order to purchase inventory.
- Mitigation measures need to be put in place to reduce the risk of customers failing to return rented products.
- It is unlikely to work in all geographies – many rural areas have low population densities, which require people to travel long distances to reach a centralised rental point, which takes time and money.
- It is usually simpler and less risky for a retailer to sell a product in one transaction than manage a rental scheme.

Pico-Solar Loans

The upfront cost barrier which prevents many customers from purchasing a pico-solar system has led to a number of microfinance (MFI) institutions developing solutions which offer energy loans to customers and also to entrepreneurs seeking to establish a business selling or renting pico-solar products. It is important to acknowledge that pico-solar systems generally involve fairly small amounts of capital which may not be as attractive a product to a bank or MFI, as higher end, more expensive, products. Any finance solution needs to take this challenge into consideration.

Marketing at the BoP

While developing ways to sell products at an affordable price is important, it is equally important to find ways to convince low income households, often located in hard-to-reach locations, to use the limited cash at their disposal to purchase a pico-solar product they may never have heard of before. Before a customer can start to want a pico-solar product and trust the company selling it, they have to know that both actually exist. The challenge is that many people simply don't know about pico-solar products, so they do not go to the store looking for them. American author Seth Godin explains in his essay, 'Marketing to the bottom of the pyramid' that 'No subsistence farmer walks to a store or stall saying, "I wonder what's new today? I wonder if there's a new way for me to solve my problems?" This is somewhat similar to the world before the motor car where Henry Ford is often quoted as saying, "If I had asked people what they wanted they would have said faster horses"'. Customers were not about to invent or imagine the car and then start spontaneously demanding it. This is called 'unarticulated need' and the challenge, therefore, is to educate potential customers, explain what a pico-solar system actually is and use marketing strategies to create demand for them.

Creating Customer Awareness

There has to be a demand for a new product or service if it is to successfully penetrate a market. Creating consumer awareness, understanding and a demand for products is, ultimately, the job of marketing.

As pico-solar products are effectively a new concept, customers need to be made aware that these products exist if they are to be expected to purchase them. The good news is that research shows that once people see and are exposed to a pico-solar product, they generally understand the value it will add to their lives. Indeed, a study carried out by Lighting Africa indicates that willingness to pay for quality pico-solar lights increases as much as five-fold with experience of using a product.

Marketing to low income households living in rural parts of the globe is quite different to marketing in more developed nations where consumers have easy access to information through television and the internet. Remote, rural areas where newspapers and televisions are a rarity are often described as 'media dark' areas. People living in these media dark areas are not only denied access to products and services due to limited infrastructure, but are also denied access to knowledge about what energy solutions exist and are available to them. The rapid rise in mobile phone ownership across the world and the corresponding improvements to communication may reduce this problem.

The challenge is to develop effective marketing strategies which can get to the hard-to-reach target market. However, accessing potential customers and educating them can be a daunting task to the uninitiated. Research conducted by Hystra explains that while 'investing in above the line marketing campaigns i.e. billboards, radio ads and TV advertising do raise awareness, they generally fail to translate into actual sales'. Instead, marketers working in these markets should 'focus their efforts on excelling at village level tactics', otherwise known as 'below the line marketing'.

In this case, below-the-line marketing can be taken to mean village level strategies which foster demand for products within the communities where the customers actually live. Examples are local product demonstrations and creating demand for and trust in products by encouraging the spread of positive messages through word of mouth from existing customers and trusted community members to potential customers.

Of course, if customers have a negative experience of a pico-solar product or company selling pico-solar, this will have a negative impact on the market and future efforts to reach customers as people spread the bad news. Figure 10.11 provides a neat summary of how word of mouth can impact sales.

Brands

As the pico-solar industry is still in its infancy, most of the leading pico-solar manufacturers and distributors are not household names and brands that are easily recognisable by consumers. As the industry grows, however, consumers will start to recognise and trust in brands which stand for quality. The mobile phone and radio industry have both been through this and many leading brands, such as Nokia and Panasonic, now have copycat brands such as 'Pana-o-sonic'.

Many people assume that low income households are not brand conscious. However, Coimbatore Prahalad (2006) argues the contrary: that 'the poor are very brand conscious. They are also extremely value conscious by necessity'. This means that they only have limited money to spend, so will often spend it carefully

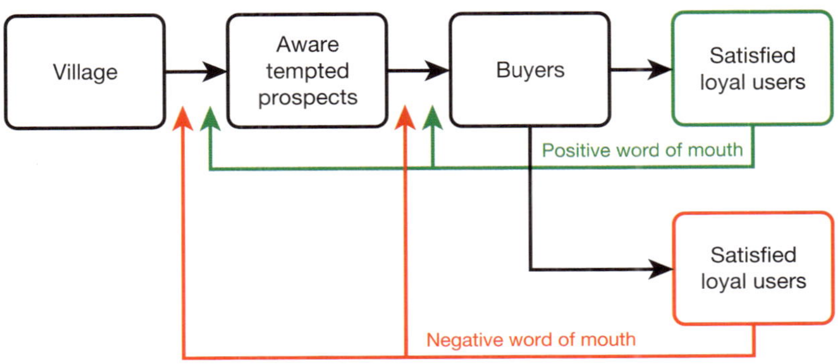

Figure 10.11 Creating demand through smart, below-the-line marketing is only part of the puzzle. Any sale needs to be backed up with a quality product and after-sales service. Only through offering a strong, dependable product and showing customers that you are there for them after a sale will they start to trust the product, trust the brand and market the product for you through 'word of mouth'. A poor experience leads to negative word of mouth.

Source: Marketing Innovative Devices for the Base of the Pyramid, Hystra 2013

and wisely. It goes without saying that a brand has to sell high quality products if it is to be trusted by the consumer. It is also important to provide the customer with a high quality of customer service.

Customer Service (After-Sales Service)

Low income households have limited money available to spend on new products and are often, therefore, far more discerning than the average consumer. Research by Hystra concludes that what deters customers from making these investments is the fear that something will go wrong and that customers will pay a higher price if they believe this will reduce the risk of product failure. It is important for customers to know that a company won't simply disappear once they have purchased a product, but instead will provide after-sales services and ensure that the customer has access to spare parts, a repairs service, and that warranties will be honoured. Ultimately, if customers have a good experience of the product and the service provided around the product, this will encourage them to spread a positive message and feed into the word of mouth marketing which is so crucial in these markets. Marketing efforts should, therefore, shift from simply raising awareness to ensuring customer satisfaction.

Good customer service means that a business works on knowing and understanding its customers' needs and goes out of its way to meet them. The development of local, well-trained sales forces that regularly interact with customers can help achieve this. The advent of the mobile phone has made it easier for companies to contact customers living in remote areas. Many companies selling pico-solar recognise the importance of customer service and have set up customer relationship management (CRM) systems designed to help them understand, communicate with and respond to their customers. Part and parcel of any good service is ensuring that customers purchase quality products in the first place, products which will last, come with a warranty and which can be repaired should they develop faults.

Product Warranties

The length of warranty offered for any product is largely a commercial decision which needs to be made by individual manufacturers and suppliers such that it varies from product to product. While PV modules and LED lights are typically designed to operate for at least five years and often much longer, the lifespan of batteries is usually much shorter and it is not uncommon to see a product warranty excluding the battery. Until recently, it was rare to see any pico-solar lighting product with a warranty which included the battery to extend beyond 12 months, and this was invariably only offered from the point of sale at the factory of the manufacturer, as opposed to at the point of sale to the final customer.

The emergence of lithium ion batteries, which are designed to last up to five years and suffer from low rates of self-discharge when not in use, has enabled companies to start offering longer warranties of 24 months to the final consumer.

Not all product problems are warranty problems, with many issues arising due to misuse and every day wear and tear. It is therefore important for all retailers to have a system to enable them to ascertain whether any fault occurring is covered by the warranty. Product manufacturers in turn need to provide distributors and retailers with training and support in this area.

Customers who live in remote parts of the world are invariably at a disadvantage compared to those in more accessible areas when they experience a product failure, due to the distances and costs involved in reaching the point of sale. Retailers selling products in these remote markets therefore need to carefully assess how customers can benefit from warranties and, where possible, implement systems to facilitate the process.

Training, Repairs and Spare Parts

While more and more products are being designed to last for many years, even the most robust products fail eventually and will need to be sent for repair if they are to remain in use. Traditionally, the battery has been the weakest part of a pico-solar system and all that is needed to extend the system's life is a replacement battery.

The challenge, however, is that batteries come in a wide range of different shapes, sizes, capacities and chemistries. This means that batteries are often not interchangeable and usually need to be replaced with the exact same battery type. This presents a challenge for technicians who effectively need to stock specialised batteries for all the different pico-solar products in the market in order to be in a position to repair them all. On a positive note, batteries are now being developed to last longer and longer and it is not inconceivable that, in the near future, batteries will be capable of operating as long as other key system component parts, such as the solar panel itself.

Whatever the product issue, however, be it the battery, circuitry, LED degradation or loose wires in the solar module, technicians need to be trained to identify and repair problems – which means access to knowledge and spare parts. One leading pico-solar distributor in Ethiopia explains that they will only supply their products to markets in other countries if they can be sure that the right systems and infrastructure are in place to ensure customers have access to quality customer care, spare parts and trained technicians. More information on product maintenance and repair is provided in Chapter 8.

Product Quality

As in any market, there are high quality products and poor quality products. There can also be fake products or products which 'oversell' themselves by claiming to have more functions or more power output than they actually do. Pico-solar markets can, therefore, be at risk of being spoiled and given a bad name. There are a number of solutions to these problems, such as:

- International standards and checks designed to protect the consumer and prevent the importation of substandard products into local markets. Chapter 7 identifies a range of standard and quality marks the consumer can look for to help them choose quality products. Many countries do not benefit from strong standards agencies, however, thus it is still possible for extremely poor quality goods to enter the market – especially in parts of Africa.

Figure 10.12 This photo shows a mobile phone repairman in Malawi opening up a faulty pico-solar light. One of the key challenges facing any skilled repairman, however, is access to the correct replacement parts.

Source: SolarAid

- Development of strong brands which customers know they can rely on, which make accurate product claims, offer strong levels of customer service and honour warranties.

Import Taxes and Tariffs

As tax and duty levels vary from country to country across the world, there is no single, straightforward answer to how much tax or duty pico-solar products attract, other than 'it depends on which country you are operating in'.

Some governments, such as Tanzania, have given pico-solar lanterns 'duty free' status and exempted them from tax. However, this is currently more the exception than the rule, with the majority of countries applying both import duty and VAT to products, thereby significantly increasing the cost of imported goods. It is also worth noting that even in countries where favourable tax conditions exist, there are many reports of inconsistent treatment of pico-solar products by local revenue authorities, charging import duty on some consignments and waiving duty on others.

The best advice to anyone interested in importing pico-solar products into a country is to seek up-to-date information from the relevant revenue authority regarding any applicable taxes, tariffs and exemptions, as situations can change quickly. If in doubt, engage a local clearing agent who should be able to advise you of any costs and any other import requirement, such as whether the shipment of goods needs a pre-shipment verification certificate.

Product Disposal and Recycling

Many countries across the world do not have sophisticated waste collection and recycling systems and infrastructure. This means that the increased use of electronic appliances is resulting in a growing global e-waste problem as products reach the end of their lifespan. While pico-solar products are generally designed to last for many years, there is a risk that these too can ultimately end up littering and polluting local environments. There are a number of things which can be done to tackle this problem:

- **Increase product lifespans:** Pico-solar products should be designed so as to ensure that batteries, which are typically the component with the shortest lifespan, can be easily replaced. This effectively increases the product's overall lifespan.
- **Ensure access to replacement batteries**: Retailers and distributors need to develop systems which enable customers to access replacement batteries or technicians who can replace the batteries for them.
- **Eco-design:** Manufactures should look for innovative ways to design products so as to minimise environmental impact of product when it reaches the end of its life, i.e. where possible, using biodegradable, reusable or recyclable, non-toxic materials.
- **National legislation**: Introduce legislation to put a framework in place which encourages development of product collection, disposal, reuse and recycling infrastructure.
- **Product collection infrastructure:** Customers need to have access to a place where they can dispose of systems responsibly and have some sort of incentive to use this facility. Many customers, especially those living in remote areas, currently have limited options in how to dispose of a product when it is no longer of use. This means the product will eventually litter the immediate environs of where it was used. Product collection costs money, however, especially in remote rural areas, thus it either needs to be subsidised, carried out or enforced by the government, or it needs to generate a revenue (for example, through product reuse or recycling) which incentivises the private

sector to carry out collection. It is unlikely to be cost effective for the private sector to collect products from expensive-to-reach rural areas unless a significant volume of products are being handled – in which case, any system needs to be well organised in order to be viable and likely to include an incentive for the customer.

- **Product disposal, recycling and reuse:** While many products which reach the end of their life are destined to end up either as landfill in official designated sites, or dumped locally as litter or in unofficial landfill sites it is possible both to dismantle products and reuse many of the components and also to recycle many of the materials.

There needs to be some sort of economic incentive for the private sector to reuse or recycle products. As the pico-solar market grows, it is likely that it will follow a similar path to the mobile phone, which has led to the rise of local repair stations where technicians make use of old products for spares and repairs. Governments and other interested parties can contribute to the evolution of this sector by offering skills training for technicians.

Product recycling requires specialised facilities which are capable of using materials such as metals and plastics. Certain volumes of materials are needed in order for recycling to be viable. The value of materials also fluctuates, and this typically requires specialist e-waste companies to enter a market. As with all markets, governments can play an important role here by incentivising e-waste companies to operate in and serve local populations.

Recognising the problem of e-waste, GOGLA, the Global Off-Grid Lighting Association (http://globaloff-gridlightingassociation.org), has set up a working group of industry actors to recommend long term solutions.

Policy and Market Facilitation – The Role of Government

This book advocates the development of a strong and vibrant pico-solar industry in which private companies meet the needs of customers, just as mobile phone companies serve customers across the world. There is a great deal that governments and other actors can do to help facilitate market development. There is also much that can be done to harm it – such as giving products away for free, which can destroy local markets for companies and entrepreneurs trying to make a living by supplying the market.

Pico-solar products, in particular lanterns and phone chargers, represent a new category of solar and should be considered to be consumer products which have been designed to meet the needs of low income households. They are, therefore, quite different to larger off-grid solar systems and should be treated differently. Many governments are still learning about these differences and the direct and positive impact pico-solar can have on people living at the BoP. Table 10.2 summarises some recommend do's and don'ts for government policy makers that want to increase access to pico-solar products in their country.

Governments can do a great deal to increase the uptake of pico-solar products. Governments can, for example, support consumer awareness campaigns, take steps to remove kerosene subsidies where these are in place and introduce more

Table 10.2 Recommendations as to how government can support the pico-solar market

	Recommendations	Comments
DO	Take steps to remove any subsidies on kerosene	Kerosene use for lighting is dangerous and should not be encouraged
DO	Provide clear guidance to companies and revenue authorities on how pico-solar systems should be classified for tax purposes	Pico-solar products are inconsistently categorised by revenue authorities, with some imports being subjected to higher taxes than others on arrival in country, creating uncertainty for companies
DO	Consider reducing taxes and tariffs for pico-solar systems	Across the board reductions can boost a market, attracting more suppliers and distributors and enabling customers to purchase systems at more affordable prices
DON'T	Reduce taxes and tariffs for some companies or for small geographic areas in a country, but not for others	Uneven, ill thought through, reductions can cause market distortion and create an uneven playing field
DO	Support schemes offering training for pico-solar technicians	This can help create skills and employment in a growing sector
DO	Establish e-waste guidelines and standards to help development of product recycling and responsible disposal infrastructure	Countries such as Kenya have developed a framework aimed at reducing the likelihood of e-waste polluting the environment
DO	Support consumer education and awareness campaigns about the benefits of solar versus fuels such as kerosene	
DO	Set up a strong quality assurance system to reduce number of poor quality products entering the market	It is important to protect the consumer from poor quality product which can harm the reputation of pico-solar
DO	Support research into the impact pico-solar systems (in particular, lighting) can have on poverty reduction	
DO	Actively engage with industry bodies and companies already working in the sector to establish what is needed to facilitate market development	While pico-solar is new to many governments, there is an increasing amount of knowledge and experience out there which can help governments make informed decisions
DON'T	Give away products for free at scale without carefully considering possible negative consequences	There is a risk that giving products away for free could harm the market for those trying to sell lights and could result in people valuing them less. Think of the mobile phone industry as an example, which has grown quickly without governments or NGOs handing out free handsets

favourable tax and tariff conditions for pico-solar systems. The introduction of more favourable tax and tariff conditions will attract more suppliers to the market and enable potential customers to access products at more affordable prices. Any changes to taxes and tariffs should, however, be implemented with care across the board and be available to everyone. If tax and tariffs are only made available for certain geographical areas in a country and not for others, this can create market distortions and ultimately cause more harm than good.

11

Case Studies

Over one-quarter of the world's population lives without regular access to electricity, including an estimated 800 million people across Asia and almost 600 million across Africa. This chapter includes a selection of four case studies from each continent where pico-solar products are being deployed by innovative, often socially driven, companies to increase access to electricity and reduce reliance on fuels such as kerosene for lighting at the base of the pyramid.

Figure 11.1 This chapter includes eight case studies from (1) East and Southern Africa, (2) Mozambique, (3) India, (4) West Africa, (5) Cambodia and (6) the Philippines.

Source: John Keane

Case Studies from Africa

Case Study: SunnyMoney – Reaching Scale in East and Southern Africa

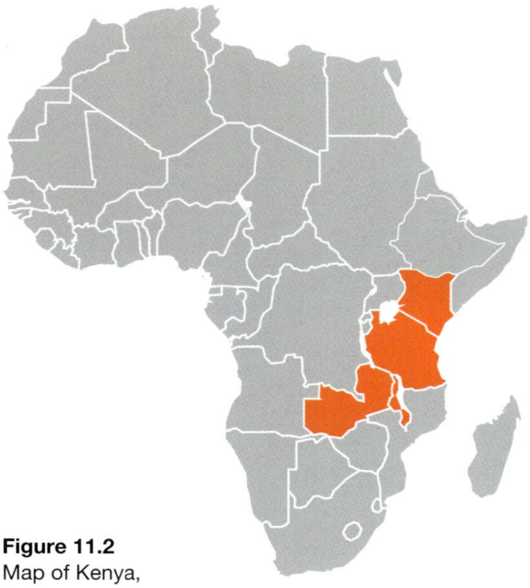

Figure 11.2 Map of Kenya, Tanzania, Malawi and Zambia

Source: John Keane

There are an estimated 110 million off-grid households across Africa which do not benefit from basic access to electricity. SunnyMoney is a social enterprise, set up and owned by the charity SolarAid, which is working towards a mission, its big hairy audacious goal (BHAG), to eradicate the kerosene light from Africa by the end of the decade (31 December 2019). The company prides itself on being a product neutral retailer which offers customers the opportunity to choose pico-solar products from a range of quality brands.

SunnyMoney has established operations in Tanzania, Kenya, Malawi and Zambia, where access to electricity, particularly in rural areas, is as low as 2 per cent and there are an estimated 80 million people forced to rely on fuels such as kerosene lights and candles each night. While there is clearly a need for alternative energy solutions, SunnyMoney believes that in order for the pico-solar market to thrive, there is a need to:

- create awareness and demand for pico-solar products;
- create trust in quality products and brands;
- overcome financial barriers;
- ensure reliable supply of products.

Following years of trialling different sales strategies, many of which were unsuccessful, SunnyMoney has developed a business model which is now enabling it to sell lights at scale in rural areas. Results to date have been impressive with sales rising from under 10,000 lights across four countries in 2010, to over 750,000 in 2013. This aggressive rise in the sales trajectory has made SunnyMoney a leading direct seller of pico-solar lights in Africa in a very short space of time. It is projecting millions of sales in the coming years.

How is This Rapid Growth Being Achieved?

SunnyMoney is achieving this rapid growth in sales through a sales model which offers entry level pico-solar lights, retailing at less than USD 10, for sale to students, enabling them to study during the evening. Education authorities support this work as they immediately grasp the potential positive impact these lights can have on education and exam results.

SunnyMoney's approach is proving so popular that it is becoming increasingly common for its rural-based teams to sell over 2000 lights in a single day. The overall strategy is to create momentum within the market by injecting

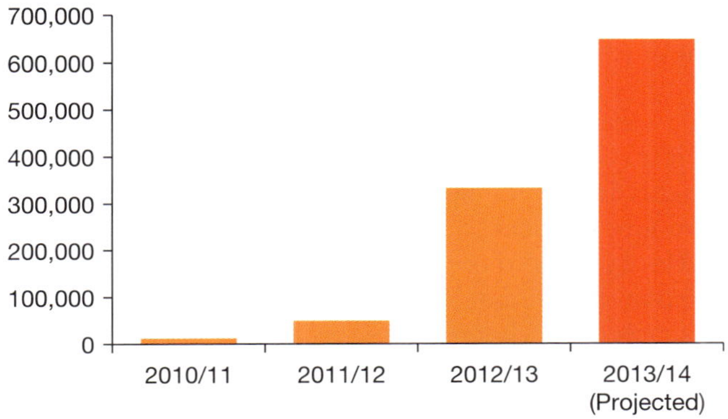

Figure 11.3 The graph shows the growth of pico-solar units sold by SunnyMoney, which has seen an explosive growth in pico-solar light sales between 2010 and 2013. This trend is set to continue as it works to eradicate the kerosene light from Africa by 2020.

Source: John Keane

Figure 11.4 Student pico-solar study light promotion leaflet in the Swahili language, explaining the many benefits of owning a light, such as time for study, saving money and improved health and safety.

Source: John Keane

thousands of entry level pico-solar lights into communities in a short space of time, thereby exposing thousands of households to the power of pico-solar, literally overnight.

The volumes help create awareness and trust in pico-solar products, educating teachers, parents and the wider community about how they work, how long they will last and what their benefits are. By selling thousands of lights, SunnyMoney is also able to achieve economies of scale and offer lights at more affordable prices. In doing so, SunnyMoney aims to create a wider demand for pico-solar lights and reach a tipping point within the market, which will ultimately make it easier to sell products through other channels, such as brick and mortar retail outlets, door-to-door salesmen and mail order.

To help it develop ongoing relationships with its customers, deliver after-sales service and drive further sales through schools and other distribution channels, SunnyMoney has developed a customer relationship management (CRM) system. One key function of the CRM system is to run targeted marketing campaigns to existing customers by sending a text message which encourages them to purchase the latest products on the market. The company believes that by helping students and households make the first step on the renewable energy ladder through the purchase of a sub-USD 10 pico-solar light, it can then introduce larger, brighter, more multi-functional products to customers over time.

By continually evolving its business, working hard to understand its customers and focusing on strong levels of customer service, SunnyMoney expects to see its sales rise year on year for the foreseeable future and expand into more markets across the African continent. It also expects the number of companies operating in the market to continue to grow as more customers make the transition to pico-solar lighting.

Figure 11.5 SunnyMoney staff take orders for solar study lights with teachers at a school in Zambia.

Source: © Steve Woodward

Figure 11.6 The head teacher of Chankhanda school, in Mzigawa, rural Zambia, with a consignment of pico-solar lights packed onto the back of a motorbike. Many teachers recognise the importance of light for education and see it as their duty as a teacher to bring lights to students. Chinese-made motorbikes are becoming increasingly affordable and common as a mode of transport across rural Africa, making it easier to travel to more remote areas which were previously reached by foot or bicycle.

Source: © Steve Woodward

Case Study: fosera – Manufacturing in Mozambique

It is estimated that less than 12 per cent of Mozambique's 23.5 million people enjoy access to electricity. Households which are connected to the electricity grid in rural areas are subjected to frequent power cuts each month, forcing the majority of the population to rely on alternative solutions for basic lighting and energy needs.

Recognising the potential market for pico-solar systems in the country, in 2012, the German company fosera set up a local manufacturing facility in the capital Maputo to produce a range of pico-solar products for sale in the local market. While the majority of pico-solar products on the global market are produced in Asia, the number of such facilities on the African continent is minimal. Establishing a local manufacturing unit enables the company to:

Figure 11.7 Map of Mozambique

Source: John Keane

- manufacture products at competitive prices comparable to the prices of products it manufactures in Asia;
- create a skilled local workforce which can repair products and enable easy access to spares;
- offer a three-year warranty on the products it manufactures;
- respond quickly to local requirements and increase production of certain product lines in response to market demand.

As a pioneering pico-solar company establishing a commercial manufacturing unit in Mozambique, fosera has inevitably faced some challenges:

- **Taxes and Tariffs:** The enterprise is designed to benefit the local economy, environment and society, such that when the company made the decision to establish the manufacturing unit in Maputo, it received assurances that the components it needs to import to enable local manufacture would benefit from preferential taxes and tariffs. fosera is yet to benefit from preferential taxes and tariffs, however. If it does secure these, it will be in a position to pass savings on to the customers and sell products at more competitive prices.

Figure 11.8 fosera staff are manufacturing a range of pico-solar lights and chargers in Maputo. The team is able to assemble products to meet demand and carry out product repairs.

Source: © fosera

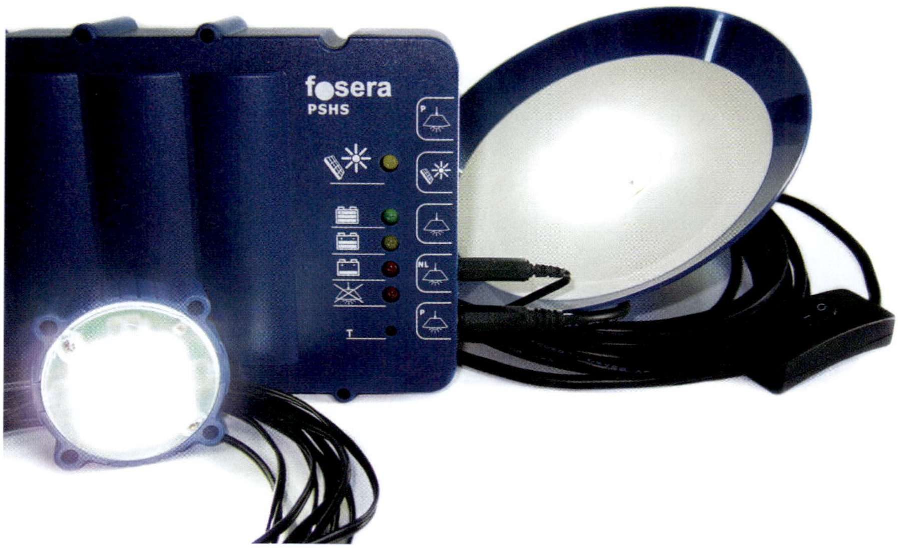

Figure 11.9 One of fosera's multi-light units assembled in Mozambique which is proving popular. The products come with a three-year warranty.

Source: © fosera

Notwithstanding this challenge, the company is still able to produce units at prices competitive to the landed cost of the goods it manufactures in Thailand.
- **Local Infrastructure:** The manufacturing unit is well positioned to produce goods for the Mozambican market. The company has found, however, that due to the costs of transporting goods across borders in Africa and export taxes out of Mozambique, it is often cheaper to send products to other countries in the region from its manufacturing base in Thailand rather than from the Mozambican unit.
- **Working Capital:** In order to realise its potential and scale up production, the company needs increased access to working capital.

Despite these challenges, the company currently employs eight staff in its factory and has produced over 5000 products for sale through retail channels in the local market. It plans to produce over 30,000 units by 2015. fosera has established a similar manufacturing unit in India and is currently looking at setting up operations in Ethiopia.

Case Study: Toyola Energy Limited – Ghana, West Africa

Figure 11.10
Map of Ghana, West Africa
Source: John Keane

Ghana in West Africa has a population of 25 million people with an estimated 9.4 million living without access to electricity. In rural areas, less than 25 per cent of the population has access to electricity. The electricity grid itself suffers from frequent power cuts, with an average of 10 blackouts reported each month. The country is ranked 135 out of 187 in the United Nation's Human Development Index, with an estimated 54 per cent of its population living on less than USD 2/day. As in much of rural Africa, people living without access to mains electricity or gas are forced to use charcoal and firewood for cooking and fuels such as kerosene and candles for lighting when it gets dark each evening just after 6pm.

Recognising that there was a real demand for electricity to power lights, charge phones and play radios, Toyola Energy Limited, a company founded in 2006 to produce and sell fuel-efficient charcoal stoves, established a solar division, Toyola Solar, which sells a range of pico-solar lighting systems. As with the stoves it retails, Toyola sells pico-solar systems through a range of channels and its own growing network of agents. Toyola have identified a number of challenges which need to be addressed in order to sell pico-solar systems successfully. The first is ensuring that potential customers trust the company selling the products. Toyola has developed a reputation for strong customer service, so customers who may otherwise be sceptical about purchasing a pico-solar light are more confident about doing so as they trust the people selling the products.

Another familiar challenge is how to overcome the upfront costs which customers face when purchasing a pico-solar product. One innovative way

Toyola helps customers overcome this barrier is through the 'Toyola Money Box' which it developed when selling stoves. The way it works is simple. During a sales promotion, customers are given a choice: they are invited to purchase a product at a special discount there and then, during the promotion, or they are given the opportunity to try the product before they buy it. In this latter scenario, however, they are not given a discount. Instead, customers are given a money box – a simple tin can – and asked to put any money they save through no longer having to charge phones, purchase kerosene or other savings, in the box.

By trying the product before buying, the customer gains confidence in the pico-solar product and the money box helps them keep track of the money they start to save – which is then used to help them afford the product. Sometimes, actually seeing the money saved is enough to convince the customer to find the money and go ahead with the purchase.

Figure 11.11 An example of a Toyola Money Box, which is a simple tin money box which enables customers and potential customers to see how much money they can save by using fuel efficient stoves and pico-solar products.

Source: © Toyola Energy

Of course, this solution has its own challenges and Toyola is careful only to provide this offer to trustworthy customers known to its local sales agents. If the customer does not want to purchase the product after the agreed trial period, the product is simply returned to the company. It is also challenging to provide this offer to too many people at any one time as it can lead to cash-flow constraints. Toyola's CEO, Suraj Wahab, believes there is a large market for pico-solar products across Ghana and neighbouring countries, and expects sales to increase each year as more customers experience the benefits of solar power.

Worldreader: Powering E-Readers in Ghana and Beyond

In much of Africa, there is a lack of access to books – UNESCO reports that there are a quarter of a billion schoolchildren in sub-Saharan Africa who have inadequate access to books of any kind. As a response, in 2010, a non-profit organization, Worldreader, introduced e-readers to rural schools in Ghana, making it possible for people living in remote locations to hold a library of books in the palm of their hand. To date, Worldreader have successfully piloted e-readers in classroom and library settings at scale across sub-Saharan Africa, delivering 441,169 books to approximately 10,000 children. Worldreader's e-reader programmes use devices that are pre-loaded with thousands of local and international textbooks and storybooks. Each device comes with a protective case, zip-around jacket, cable and battery-powered book light so the children can read after dark.

Figure 11.12 Schoolchildren in Ghana show off their e-readers which are kept charged by a pico-solar system integrated into the protective carry case.

Source: © Worldreader

While e-readers are extremely energy efficient since they use an e-ink screen, which draws less power than the LCD screen more commonly found on laptops and tablets, they still need to be recharged periodically and light is needed for night-time reading. The batteries in the devices themselves also need to be replaced every few years. For the many schools across Africa which do not have access to electricity, Worldreader decided to look into solar solutions to keep the e-readers charged.

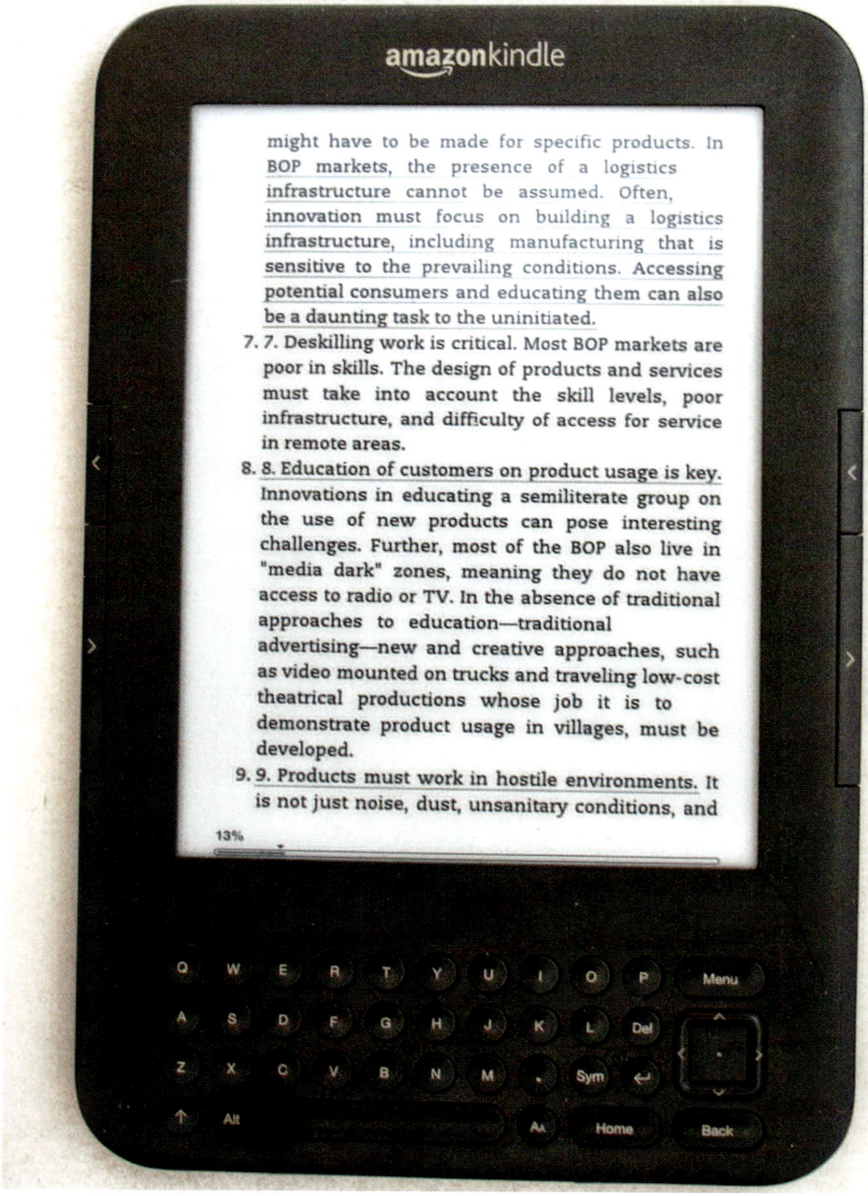

Figure 11.13
The e-readers that Worldreader uses have 150 and 300 hours of battery life, respectively, requiring charging weekly or fortnightly rather than daily. As with any rechargeable electronic device, e-readers need a new battery every few years. The e-readers themselves will also, ultimately, need to be replaced when they come to the end of their lifespan and need to be disposed of properly. While e-readers are falling in price, they are still too expensive for most people living across Africa. For this reason, the e-readers in this case study are intended for public use as opposed to private ownership.

Source: © Worldreader

Worldreader experimented with various pico-solar options, some of which were donated by generous manufacturers. In field testing, three important considerations soon emerged:

1. Some pico-solar devices came with a number of different components and connectors, which were prone to getting lost.
2. Recipients were often more excited about seeing what else their pico-solar device could charge rather than using it for the e-readers. Solar power itself is a social good, yet in an environment where Worldreader was reaching the youngest and often most disenfranchised members of society, the result was that others would take their solar chargers, leaving them unable to charge their e-readers.
3. Ease of use was an issue – a 7-year-old with a very limited command of English had to be able to charge her own e-reader.

In order to ensure that the e-readers remain operational in challenging environments, Worldreader has worked hard to ensure that all the e-readers it uses are well protected. They have learned, for example, that e-readers need a good protective case, which is:

1. Made of a material that wipes clean. Any washable material would be scrubbed to pieces by the children, who every week would take a stiff-bristled brush and a bucket full of soapy water to their e-readers, merely to keep them free of the ever present dust.
2. Able to fully enclose and protect the e-reader against any impact or pressure against the screen – the most fragile component of the e-reader.

By combining the need for a protective case with the need for power, Worldreader decided that the solution to both was to use a case with its own integrated pico-solar system. After testing an existing pico-solar charging case on the market, Worldreader discovered that the majority of its needs were already met. Specifically, the e-reader would be charged even when the solar module was in indirect sunlight. The device was also easy to use, and served a single function. However, with a price tag of about USD 100, this solution was not going to be economically viable. Testing also revealed that users preferred to use inexpensive clip-on lights as opposed to the integrated light in the case.

After talks with the manufacturer a new prototype case was developed which would cost significantly less, wipe clean, come with a cross-body strap, and contain an internal battery for charging anytime, anywhere. These prototypes are currently being tested in Ghana. Worldreader hopes that it will be able to iterate on the design and specifications of the prototype, and work towards rolling out an integrated e-reader solar charging case in all its programmes throughout sub-Saharan Africa, introducing more and more children worldwide to the power of reading.

Figure 11.14 An e-reader case incorporating a pico-solar module and internal battery is robust, easy to use and to keep clean.

Source: © Worldreader

Figure 11.15 The reader case incorporates a lithium ion battery and recharges the e-reader via a micro-USB charging cable. As e-readers are not yet common in Africa, replacement batteries for the e-readers and the solar chargers are not readily available on the market. Any project involving e-readers needs to ensure the availability of replacement batteries and train technicians so that devices can be locally repaired.

Source: © Worldreader

Case Studies from India

India is home to the largest off-grid population in the world with some 400 million people living without access to energy and a further 420 million people facing 'significant under-electrification'. This forces large proportions of the population to rely on kerosene as a fuel, which is subsidised to about 30 per cent of the market rate. This section looks at two innovative companies, Greenlight Planet and Orb Energy, working hard to provide India's off-grid population with access to clean, renewable energy.

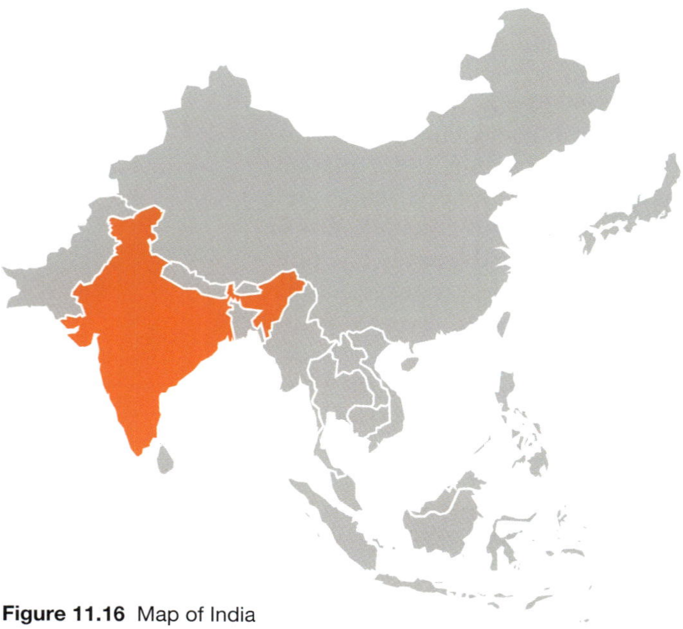

Figure 11.16 Map of India

Source: John Keane

Greenlight Planet: Direct-to-Village (DTV) Distribution

Greenlight Planet is a company which manufactured its first pico-solar light, the Sun King, in 2008. Since then, the company has developed a range of pico-solar lighting products and will soon have sold over a million units. In addition to product design and manufacture, the company has been developing a proprietary sales and distribution channel called the Direct-to-Village (DTV) model. Through this model, the company has identified and trained over two thousand rural-based sales agents called *saathis* across three states, Bihar, Orissa and Uttar Pradesh, and sold over 400,000 units to date. The company plans to increase this sales force to 15,000 *saathis* by 2015.

Figure 11.17 Point of sale material in Hindi, used by *saathis* across India to promote the SunKing Pro light and phone charger unit and the SunKing solar light.

Source: Greenlight Planet

Figures 11.18 and 11.19 Sun King pico-solar light lighting up a home, enabling chidren to study (top); Sales agents in India (saathis) wearing branded t-shirts at a team training (bottom).

Source: Greenlight Planet

Saathis tend to have a separate daytime job and sell their pico-solar lights at night as an additional source of income. *Saathis* typically boost their earnings by an additional 40–50 per cent, selling between 10 to 20 lights per month on average, with some managing to sell over 75 lights and increasing their income by as much as 100–200 per cent. Meanwhile, the payback period for customers who purchase a pico-solar light is generally 4–8 months, based on what they would typically spend on kerosene.

Saathis live in the communities they serve and are therefore at the heart of the company's sales and marketing strategy. In addition to building trust in pico-solar and the Sun King brand, they often earn tremendous respect from their communities in the process. *Saathis*, driven by their will to earn for their family, enhance their trusted word of mouth sales efforts with colourful, local language marketing materials (see Figure 11.17 on p. 159) and organise innovative branded events, creating a spectacle on consumers' doorsteps. These sales generation activities come at much lower expense, with much greater efficacy than more traditional marketing strategies.

The company holds regular training and coaching sessions for its *saathis* which focus on helping them make face-to-face sales and run product demonstrations. *Saathis* are also trained on how to deal with warranty issues and provide customer service. As sales grow and the company seeks to expand the DTV model to more districts and states, one of the biggest challenges it faces is working capital. The company is therefore putting a lot of effort into optimising its inventory management at each stock point and state level warehouses for 'just in time' levels, so as to minimise the amount of stock remaining sedentary and tying up working capital unnecessarily.

Orb Energy

Orb Energy is an award-winning solar company set up in 2006, selling a wide range of solar solutions through a network of dealers, distributors and franchisees across the Southern Indian states of Karnataka, Kerala, Maharashtra, Tamil Nadu and Andhra Pradesh. The company is convinced that there is a real need for pico-solar solutions, not only across India's unelectrified population, but also in areas where the electricity grid suffers frequent power shortages. In response to this perceived need, the company has developed a range of 'plug and play' pico-solar systems ranging from a 5 watt two light system to a 10 watt four light system. Since introducing pico-solar systems into their shops in 2012, they make up on average 15 per cent of the company's unit sales.

The company currently sells hundreds of these systems each month alongside larger solar systems through a network of dealers and distributors, with the 5 Wp system retailing at 4900 rupees (around USD 90). Orb's objective is to increase sales to tens of thousands of units each month. In order to achieve this, Orb's CEO, Damian Miller, has identified the need to create 'a pull through the distribution channel'. At the moment, many dealers need to be convinced or motivated to actively promote the sale of pico-solar systems which offer less revenue than their larger counterparts. Miller explains, 'It's sometimes hard to get dealers excited by low revenue products . . . unless they can sell a lot of them'.

Figure 11.20
Orb Energy's Solectric Plug and Play system. This multi-light pico-solar system is sold by Orb across India. The system includes a USB outlet, enabling it to charge mobile phones and other small USB-powered devices.

Source: Orb Energy

Herein lies the challenge. As pico-solar systems are new entries into the solar market, many people simply do not know that these products exist and they do not yet sell in high volumes. As a result, Miller explains, 'you simply don't get large numbers of potential customers actively walking into a shop looking for these products'.

The challenge, therefore, is to create demand and market momentum, which takes time, effort and money. In response to this, Orb has set up a dedicated team which provides its sellers with marketing and sales support, which includes important solar demonstrations in villages to raise general awareness about the existence of pico-solar systems and ultimately convince people that they want one in their homes. Through concerted marketing efforts, Orb expects to see a steady increase in sales. It plans to diversify its product range to offer a more affordable single light and phone charging unit retailing closer to USD 35 in a bid to attract households which cannot afford a USD 90 purchase without some sort of finance. Orb is also planning to expand its sales reach into the more remote unelectrified areas of west and northern India where it expects a large market for pico-solar lights and phone charges.

Case study from Cambodia

Kamworks – Creating Lighting Solutions

Figure 11.21
Map of Cambodia
Source: John Keane

In Cambodia official statistics indicate that approximately 80 per cent of the country's 14.5 million people do not have access to the electricity grid and are forced to rely on car batteries, torches, candles and kerosene for power and light.

In response to this situation, a social enterprise called Kamworks was established in 2006 to provide access to affordable sustainable energy systems. One way Kamworks is working to achieve this is through the development of a pico-solar lighting product called the MoonLight which has been designed to offer a substitute to the kerosene lamp. Following extensive field research, Kamworks established how lighting is currently used in rural Cambodia.

Kamworks market the light to all sections of society, although the greatest need is in the rural areas where Kamworks is trying to reach through a range of distribution partners. The light retails for less than USD 20, a price which is more competitive than many other pico-solar lights, but still a more expensive upfront cost than a kerosene lantern. The payback period is estimated to be less than 12 months as customers make savings by no longer having to purchase kerosene fuel. Recognising upfront costs as a barrier to purchase, Kamworks has trialled a rental model whereby village entrepreneurs access a loan from local microfinance organisations to purchase multiple lights which are then rented out to villagers at costs competitive to daily kerosene use. At its peak the company had helped set up 100 village entrepreneurs renting out 2000 MoonLights. While this model does overcome financial barriers for customers, it requires significant organisation. To date the company has sold more than 20,000 MoonLights across Cambodia and is expecting sales to increase year on year.

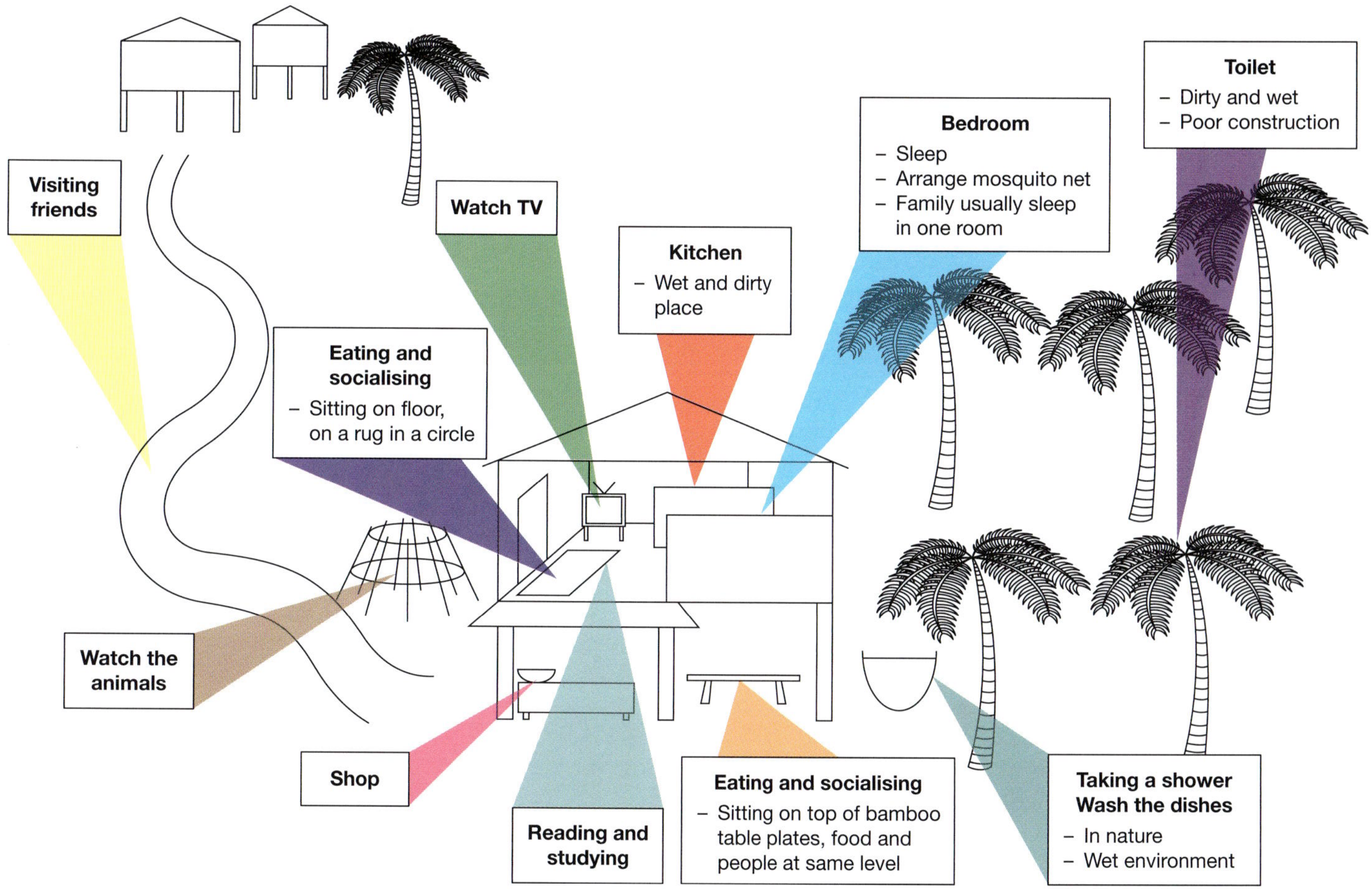

Figure 11.22 Research findings illustrating the many different situations where light and electricity is needed during the evening in Cambodia.

Source: Kamworks

Figure 11.23 The MoonLight was designed following extensive field tests and research in Cambodia. It is both a light and a phone charger.

Source: Kamworks

Figure 11.24 Children using the MoonLight to read after dark in Cambodia.

Source: Kamworks

Figure 11.25
Map of the Philippines
Source: John Keane

Case Study from the Philippines

In the Philippines, access to grid electricity is relatively high at 86 per cent, but reliability is poor, with frequent blackouts. There are also still some 2.5 million households without access to electricity, the majority of these located in rural areas.

The Negros Women for Tomorrow Foundation (NWTF)

The Negros Women for Tomorrow Foundation (NWTF), a microfinance institution, has developed a range of finance solutions and loans aimed at enabling more people to access pico-solar products. Serving a base of 140,000 clients, NWTF offers energy loans of between USD 35 and USD 200 repayable over 6 to 12 month periods with a flat interest rate of 2.5 per cent per month. These loans enable customers to purchase a range of pico-solar products from single lighting to multi-light systems and immediately start benefiting from savings made through cuts of between 30 to 40 per cent to their energy budget. Most customers make their final repayment for a product within 12 months.

Since establishing the pico-solar initiative in 2009, over 6000 pico-solar lights have been sold through 100 sales agents. During this time, NWTF has evolved its operations to overcome a number of challenges.

Figure 11.26 Entrepreneurs in the Philippines are selling pico-solar lights and phone chargers alongside other retail goods in local shop outlets.

Source: NWTF

Multiple Loans

To avoid clients being forced to choose between an energy loan or loan for some other need, NWTF enables clients to take out more than one loan at a time, depending on their credit and repayment history. Offering more than one loan at one time helps overcome the challenge which faces many MFIs whereby the overhead associated with offering the relatively small amount of money needed for a pico-solar loan is shared across multiple loans.

Incentivising Sales

The relatively small amount of money involved in pico-solar loans means that there can be less incentive for a loan officer to promote such a loan if they are incentivised by the value of loans they are responsible for. NWTF have overcome this issue by offering a flat commission to their loan officers (the value varying depending on product). NWTF has also introduced a support system of Green Energy Officers whose role it is to support its loan officers, helping to market and explain the benefits of pico-solar products to customers. In a bid to attract more pico-solar customers, NWTF has created a related loan initiative which enables its clients to become energy sales agents who can sell products to the wider market. NWTF structures the loan in such a way that energy sales agents can in turn extend small amounts of credit to customers in their communities.

Customer Default

While customer defaults are low across NWTF's portfolio at less than 3 per cent, they did experience high default levels of up to 25 per cent with an early pico-solar product which suffered from a range of technical faults. Ensuring products are of a high quality with minimal faults minimises the risk of financial defaults. Raymond Serios, NWTF's Energy Programme Director, explains that defaults for pico-solar systems are now at 1.5 per cent and they expect to see annual light sales to reach 5000, with the most popular pico-solar systems offering customers 2–4 lighting points.

The pico-solar market in the Philippines is set for a further boost in 2013 with the energy giant, Total, planning to stock products for sale at its network of gas stations, and pioneering solar company Suntransfer planning to set up a network of pico-solar retail and service stations in the country.

12

Resources

This chapter provides resources and suggested reading for people interested in finding out more about the off-grid solar and pico-solar industry. It includes details of key references used in writing this book and provides links to a selection of industry bodies, pico-solar companies, programmes, forums and initiatives aimed at developing the off-grid energy sector.

Further Reading

Hankins, M, *Stand-Alone Solar Electric Systems: The Earthscan Expert Handbook for Planning, Design and Installation*, 2010

Lighting Africa, *Market Trends Report – Overview of the Off-Grid Lighting Market in Africa,* Lighting Africa, 2012

Lighting Asia, *Solar Off-Grid Lighting: Market Analysis of India, Bangladesh, Nepal, Pakistan, Indonesia, Cambodia and Philippines*, 2012

Lighting Global Technical and Eco Design Notes Series: www.lightingafrica.org.

Lysen, E., *Pico Solar PV Systems for Remote Homes: A New Generation of Small PV Systems for Lighting and Communication*. IEA PVPS Task 9 Report IEA-PVPS T9-12, 2012

Miller, D., *Selling Solar: The Diffusion of Renewable Energy in Emerging Markets*, Earthscan 2010

Industry Bodies, Initiatives and Programmes

Lighting Africa, Lighting Asia and Lighting Global

Lighting Africa, Lighting Asia and Lighting Global are joint World Bank/IFC programmes, which aim to improve access to clean, affordable lighting, particularly in Africa and Asia. The programmes exist to catalyse and accelerate the development of sustainable markets for affordable, modern off-grid lighting solutions for low income households and small enterprises.

The overall approach is to accelerate the development of off-grid lighting markets by:

- demonstrating the viability of the market to companies and investors by providing market intelligence on market size, consumer preferences and behaviour, and on the base of the pyramid business models and distribution channels;

- improving the enabling environment for the sector by developing quality assurance market infrastructure, and facilitating business-to-business interactions through conferences, workshops and a dedicated web platform. The programmes also work with governments to address policy barriers;
- supporting the scale up and replication of successful businesses by providing targeted business development services and facilitating access to finance for manufacturers, local distributors and other stakeholders.

Lighting Global produces a series of useful technical and eco design notes for products and carries out non regional specific activities to support Lighting Africa, Lighting Asia and the wider market.

Website: www.lightingafrica.org

LuminaNET

LuminaNET is a social network for the global off-grid lighting community. Developed by the Lumina Project, based at the Lawrence Berkeley National Laboratory, the network is dedicated to solving problems arising from fuel-based lighting in the developing world. Its purpose is to amplify the collective knowledge base by offering a space for sharing of knowledge, discussing emerging technologies and business models, and joint problem-solving. LuminaNET offers blogs from community members, an interactive discussion forum, a calendar of industry events, photos, videos, a marketplace with information about products and services, and a member-generated directory of off-grid lighting field projects.

Website: http://luminanet.org/

Global Off-Grid Lighting Association

The Global Off-Grid Lighting Association (GOGLA) has been established to act as the industry advocate with a focus on small and medium enterprises. It is a neutral, independent, not-for-profit association created to promote lighting solutions that benefit society and businesses in developing and emerging markets. GOGLA will support industry in the market penetration of clean, quality alternative lighting systems. Formed in 2012 as a public–private initiative, GOGLA was conceived out of the joint World Bank/IFC effort to provide a sustainable exit strategy for Lighting Africa initiative. GOGLA has set up Working Groups to develop guidance and recommendations for the following areas:

- Policy and regulation
- Quality and standards
- Life-cycle and recycling
- Business models and market intelligence

Website: http://globaloff-gridlightingassociation.org/

Global LEAP

Launched in 2012, Global LEAP's mission is to drive and support sustainable commercial markets that increase modern energy access worldwide. The programme's structure includes activities in five main areas: product quality assurance, finance across the supply chain, market intelligence, consumer education and policy.

Website: http://www.globalleapawards.org/

Sustainable Energy for All

Sustainable Energy for All is an initiative launched by the United Nations Secretary-General to make sustainable energy for all a reality by 2030. It has three key objectives:

1. Ensure universal access to modern energy services.
2. Double the global rate of improvement in energy efficiency
3. Double the share of renewable energy in the global energy mix.

An Energy Access Practitioner Network has been set up as part of the Sustainable Energy for All Initiative. The Network focuses on household and community-level electrification for productive purposes, incorporating specific market-based applications for health, agriculture, education, small business, communities and household solutions.

Website: http://www.sustainableenergyforall.org/

sun-connect

sun-connect is an information service focusing on rural development through solar energy. It circulates a monthly newsletter and shares information through its website. The target audience includes solar firms, micro-credit organizations, NGOs and state facilities in developing and emerging nations.

Wesbite: http://sun-connect.org

International Electrotechnical Commission (IEC)

The IEC is an international standards organisation that prepares and publishes International Standards and guidelines for electrical, electronic and related technologies. IEC/Technical Specification 62257-9-5:2013(E) applies to stand-alone rechargeable electric lighting appliances or kits that can be installed by a typical user without employing a technician. This technical specification presents a quality assurance framework that includes product specifications and test methods.

Website http://www.iec.ch

Selection of Pico-Solar Companies Specialising in Lighting Products for the Base of the Pyramid (BoP) Market

This selection of companies is provided for information purposes only. It is not intended to be a comprehensive list.

Table 12.1

Barefoot Power barefootpower.com Manufacturer	**Solux Service GmbH** www.solux.org Manufacturer	**Schneider Electric** schneider-electric.com Manufacturer	**Shanghai Roy Solar** www.roysolar.com Manufacturer
d.light design dlightdesign.com Manufacturer and Distribution	**fosera** www.fosera.com Manufacturer	**Nokero** www.nokero.com Manufacturer	**Sunlite** www.sunlite.co.ke Manufacturer
Greenlight Planet greenlightplanet.com Manufacturer and Distribution	**Kamworks** www.kamworks.com Manufacturer and Distribution	**Thrive Solar Energy Ltd** www.thriveenergy.co.in Manufacturer and Distribution	**SunnyMoney** www.sunnymoney.org Distribution
Solar Sisters www.solarsister.org Distribution	**Orb** www.orbenergy.com Manufacturer and Distribution	**One Degree Solar** onedegreesolar.com Manufacturer	**Little Sun** www.littlesun.com Manufacturer
Omnivoltaic www.omnivoltaic.com Manufacturer	**Pharos Off Grid** www.pharosoffgrid.com Manufacturer	**Angaza Design** angazadesign.com PAYG Manufacturer	**Azuri** azuri-technologies.com PAYG Manufacturer
Betta Lights www.bettalights.com Manufacturer	**Philips** www.philips.com Manufacturer	**Trony** www.trony.com Manufacturer	**SolarAid** www.solar-aid.org Charity
Waka Waka www.waka-waka.com Manufacturer			

Bibliography

Adelman, P. and Reindl, T. (2012). *Batteries for Pico PV Systems*, Solar Energy Research Institute of Singapore (SERIS), National University of Singapore 3rd Lighting Africa Conference.

Agrawal, R. and Dutt, N. (2013). *Distribution Channels to the Base of the Pyramid*, Business Innovation Facility Issue 5.

Alstone P., Mills E. and Jacobson A. (2011). *Embodied Energy and Off-Grid Lighting*, The Lumina Project, Technical Report no. 9.

Bandura, A. (1993). 'Perceived Self-Efficacy in Cognitive Development and Functioning', *Educational Psychologist*, 28:2 (1993), pp. 117–148.

Battery University, http://batteryuniversity.com/learn/article/memory_myth_or_fact

BBC Future (International version). (2012). 'Charging Tomorrow's Smartphones', http://www.bbc.com/future/story/20120227-charging-tomorrows-smartphones/1

Bond, T. C., Zarzycki, C., Flanner, M. G. and Kock, D. M. (2011). 'Quantifying Immediate Radiative Forcing by Black Carbon and Organic Matter with the Specific Forcing Pulse', *Atmospheric Chemistry and Physics*, 11, 1505–25.

BuildingTechnologies Program (2013). US Department of Energy, *LED Basics*, Energy Efficiency and Renewable Energy, http://www1.eere.energy.gov/buildings/ssl/sslbasics_ledbasics.html

City of Portland, Oregon (2011). *Solar-Powered SmartMeters Streamline Portland's Parking*, http://www.portlandoregon.gov/bibs/article/157993

d.light (2013). at http://www.dlightdesign.com/impact-dashboard/customer-benefits/

Eco Design Notes (2012). *Battery Toxicity and Eco Product Design*, Lighting Global, Issue 1, September.

Esper, H., London, T. and Kanchwala, Y. (2013). *Access to Clean Lighting and their Impact on Children: An Exploration of SolarAid's SunnyMoney*, Child Impact Case Study No. 4. Ann Arbor: The William Davidson Institute.

Godin, S. (2010). 'Marketing to the Bottom of the Pyramid,' http://sethgodin.typepad.com/seths_blog/2010/09/marketing-to-the-bottom-of-the-pyramid

Government of India, Ministry of Home Affairs (2011). http://www.censusindia.gov.in/2011-prov-results/indiaatglance.html

Government of India (2011). Census of India: *Houses, Household Amenities and Assets*.

Gruner, R., Lux, S., Reiche, K. and Schmitz-Gunther, T. (2009). *Solar Lanterns Test: Shades of Light*, Eschborn: GTZ, www.gtz.de

GTZ (2010). *What Difference Can a PicoPV System Make? Early Findings on Small Photovoltaic Systems – An Emerging Lowcost Energy Technology for Developing Countries* Eschborn: www.gtz.de

Guidelines to Defra/DECC's GHG Conversion Factors for Company Reporting (2011). Produced by AEA for the Department of Energy and Climate Change (DECC) and the Department for Environment, Food and Rural Affairs (Defra).

Guryan, J. (2008). 'Parental Education and Parental Time with Children', *Journal of Economic Perspectives, American Economic Association*, 22:3, 23–46.

Hankins, M. (2010). *Stand-Alone Solar Electric Systems: The Earthscan Expert Handbook for Planning, Design and Installation*, London: Routledge.

Hystra Hybrid Strategies Consulting (2013). *Marketing Innovative Devices for the Base of the Pyramid*, http://hystra.com/marketing-devices

Kammen, D. (2011). Chief Technical Specialist for Renewable Energy and Energy Efficiency at the World Bank, The Institute of Science in Society (ISIS) (2011). *ISIS Report 06/04/11*, http://www.i-sis.org.uk/LightingAfrica.php?printing=yes

Lam, N., Smith K., Gauthier, A. and Bates, M. (2012). 'Kerosene: A Review of Household Uses and their Hazards in Low- and Middle-Income Countries', *Journal of Toxicology and Environmental Health*, Part B: Critical Reviews, 15:6, 396–432.

Lam, N. L., Chen, Y., Weyant, C., Venkataraman, C., Sadavarte, P., Johnson, M. A. and Bond, T. C. (2012). 'Household Light Makes Global Heat: High Black Carbon Emissions from Kerosene Wick Lamps', *Environmental Science & Technology*, 46:24, 13531–8.

Lighting Africa (2010). *Solar Lighting for the Base of the Pyramid – Overview of an Emerging Market*, Lighting Africa.

Lighting Africa (2012). *Market Trends Report – Overview of the Off-Grid Lighting Market in Africa*, Lighting Africa.

Lighting Asia (2012). *Solar Off-Grid Lighting: Market Analysis of India, Bangladesh, Nepal, Pakistan, Indonesia, Cambodia and Philippines*, Lighting Asia 2012.

Lighting Global Technical and Eco Design Notes Series, (2010–). at www.lightingafrica.org

Lysen, E. (2012). *Pico Solar PV Systems for Remote Homes: A New Generation of Small PV Systems for Lighting and Communication*, IEA PVPS Task 9 Report IEA-PVPS T9-12.

Miller, D., (2010). *Selling Solar: The Diffusion of Renewable Energy in Emerging Markets*, London: Earthscan.

Mills, E. (2005). 'The Spector of Fuel-Based Lighting', *Science AAAS*, 308, http://light.lbl.gov/pubs/mills_science_fbl.pdf

Narasimha Desirazu Rao (2011). *Distributional Impacts of Energy Policies in India: Implications for Equity*, Stanford University.

Nique, M. (2010). *Off-Grid Handset Charging, Green Power for Mobile*, Bi-Annual Report, November, http://www.gsma.com/mobilefordevelopment/wp-content/uploads/2012/04/offgridhandsetcharging.pdf

Prahalad, C. K. (2006). *The Fortune at the Bottom of the Pyramid: Eradicating Poverty through Profits*, Upper Saddle River, NJ 07458: Pearson Education Inc. publishing as Prentice Hall.

Prugue, X. (2012). *Energy Poverty: India's Best Kept Secret*, United Nations Environment Programme (UNEP), http://new.unep.org/wed/blogs/Prugue4.asp

Purdue University. (2003). 'Modeling Appropriate Behavior', *Parent Provider Partnerships*.

Reddy, T. (2010). *Linden's Handbook of Batteries*, 4th edn. New York: McGraw-Hill.

SolarAid (2012). *Data from Follow up Interviews with 232 Pico-Solar Customers in Narok County in Kenya, Karonga District in Malawi and Kilimanjaro Region in Tanzania.*

The Economist (2010). 'Power to the People: Energy in the Developing World' (2 September 2010), http://www.economist.com/node/16909923.

UN (2007). *Education is Key to Reducing Child Mortality: The Link between Maternal Health and Education*, UN Chronicle, http://www.un.org/wcm/content/site/chronicle/home/archive/issues2007/themdgsareweontrack/educationiskeytoreducingchildmortalitythelinkbetweenmaternalhealthandeducation.

UN Press Release (2013). *Sustainable Off-Grid Lighting Solutions Can Deliver Major Development and Climate Benefits*, UNEP Announces new partnership with Global Off-Grid Lighting Association, at http://www.rona.unep.org/documents/news/en%20lighten%20PR%20GC%20140213%20FINAL.pdf.

UN (2013). *A New Global Partnership: Eradicate Poverty and Transform Economies through Sustainable Development*: The Report of the High-Level Panel of Eminent Persons on the Post-2015 Development Agenda, at http://www.un.org/sg/management/pdf/HLP_P2015_Report.pdf

UNESCO (2005–2010 data). *Regional Literacy Rates for Adults (15+)*, http://stats.uis.unesco.org/unesco/TableViewer/tableView.aspx?ReportId=201.

UNESCO (2012). *Education for All Global Monitoring Report: Youth and Skills: Putting Education to Work*, http://unesdoc.unesco.org/images/0021/002180/218003e.pdf

UNFCCC 2012: AMS-III.AR (2012). *Substituting Fossil Fuel Based Lighting with LED/CFL Lighting Systems*, at http://cdm.unfccc.int/methodologies/DB/41A0Q0QT5CUP3TMD57GC6RZ4YRV28M, p. 9.

Verschelling, J. and Diehl, J. C. (2010). *Solar Lantern Development and Dissemination – From Participatory Design to Implementation in Rental Scheme*, www.kamworks.com

WBCSD Development (2013). *Business Solutions to Enable Energy Access for All*, The WBCSD Access to Energy Initiative, http://www.wbcsd.org/pages/edocument/edocumentdetails.aspx?id=14165&nosearchcontextkey=true

World Bank (2008). *The Welfare Impact of Rural Electrification: A Reassessment of the Costs and Benefits*, An IEG Impact Evaluation, http://lnweb90.worldbank.org/oed/oeddoclib.nsf/DocUNIDViewForJavaSearch/EDCCC33082FF8BEE852574EF006E5539/$file/rural_elec_full_eval.pdf

World Bank (2011). 'Early Childhood Development: Nutrition', http://go.worldbank.org/DL9AKYWQ70> in *ibid.*

World Bank, IFC, MIGA, Independent Evaluation Group (2010). 'What Can We Learn from Nutrition Impact Evaluations: Lessons from a Review to Reduce Child Malnutrition in Developing Countries', in *ibid.*

World Bank (2012). *Mobilizing the Agricultural Value Chain*, http://siteresources.worldbank.org/EXTINFORMATIONANDCOMMUNICATIONANDTECHNOLOGIES/Resources/IC4D-2012-Chapter-2.pdf

World Energy Outlook (2011). International Energy Agency, http://www.worldenergyoutlook.org/resources/energydevelopment/accesstoelectricity/

Glossary

Alternating current (AC) – electric current that reverses direction in a circuit at regular intervals
Amorphous silicon – type of thin film solar photovoltaic material
Amp – unit of electric current which measures the flow of electrons passing a point in an electric circuit per unit time
Amp-hour (Ah) – the amount of charge in a battery that will allow one amp of current to flow for one hour; used to indicate battery capacity
Appliance – electronic device which uses electricity in order to operate, such as a phone or radio
Base of the pyramid (BoP) – term used to define low income populations who typically live on less than USD 2.5/day; sometimes referred to as 'bottom of the pyramid'
Battery – an energy storage device which stores chemical energy which can be drawn out as electrical energy to power electronic device
Battery capacity – amount of electrical charge a battery is capable of storing; usually measured in amp-hours
C-rate – measure used to explain how quickly a battery is being charged or discharged
Charge controller – circuitry or device which is designed to protect rechargeable batteries by limiting the rate at which electric current is drawn from or added to rechargeable batteries, ensuring the battery is not overcharged or discharged
Circuit – closed system of wires and conductors which allow electric current to flow
Compact fluorescents (CFLs) – type of energy efficient light bulb
Charge cycle – process of charging and discharging a rechargeable battery; batteries can operate for a finite number of charge cycles; the number of charge cycles is used to explain a battery's lifespan
Depth of discharge – the amount of charge removed from a battery, expressed as a percentage of the total battery capacity
Direct current (DC) – electric current flowing in one direction; pico-solar systems typically only use direct current
Electric current – the rate of flow of electrons in a circuit
Heat sink – component, usually metal, that cools a device by dissipating heat; used to prolong LED life by preventing overheating and potential damage
Ingress protection (IP) – rating system used to explain degree to which a product is protected against the intrusion of water and solid objects and dust
I-V curve – relationship between the electric current and voltage produced by a solar cell or module – plotted on a graph; used to determine performance at different levels of insolation and temperature
Illuminance – the degree to which a surface area is illuminated; measured in lux
Insolation – measure of solar radiation energy received on a surface area and at a given time; also called solar insolation

Irradiance – solar radiation incident on a surface per unit time; measured in watt per square metre (w/m^2)
Light-emitting diode (LED) – an efficient semiconductor light source which dominates the pico-solar sector; uses include general lighting and as a charge controller indicator
Lithium ion – umbrella name used to categorise range of different lithium-based rechargeable battery chemistries
Load – appliance which uses electrical power
Low voltage disconnect – control circuit which automatically disconnects circuit when voltage falls below a set voltage threshold to prevent battery from over-discharging
Lumen – unit measure of the total amount of visible light emitted by a light source
Luminous efficacy – the efficiency with which electrical power is converted into light; measured in lumens per watt (lm/w) as the ratio of light output to power input
Luminous flux – the total amount of visible light emitted in all directions by a light source; measured in lumens
Luminous intensity – how bright a light appears to be in a particular direction; measured in candelas (cd)
Lux – a measure of the intensity of visible light that hits surface; one lux equals one lumen per square metre (l/m^2)
Maximum power point – the point at which a solar module produces the most power; clearly identifiable on an I-V curve
Micro-solar – see pico-solar
Milliamp-hour (mAh) – one thousandth of an Amp-hour (0.001Ah), commonly used to indicate capacity of pico-solar batteries
Multimeter – an instrument with multiple settings, used to measure electrical units such as voltage, current and resistance
Non-government organisation (NGO) – organisation operating independently from government, normally on a not for-profit basis
Nickel-cadmium – type of rechargeable battery
Nickel-metal hydride – type of rechargeable battery
Off-grid – term used to refer to households and communities which are not connected to the main electricity grid
Ohm – unit of resistance
Open-circuit voltage (Voc) – the maximum voltage measured across a solar module when no load is attached and no current is flowing
Peak sun hour – the equivalent number of hours each day, during daylight, when solar irradiance averages 1000 w/m^2; used to calculate system sizes as more peak sun hours mean a solar PV module can produce more electricity
Photovoltaic (PV) – refers to the conversion of sunlight (solar radiation) into electric current
Pico-solar – term used to define products and systems which utilise small solar pv modules, typically below 10 W peak
PicoPV – see pico-solar
Potential difference (voltage) – the potential difference in energy between two points in a circuit which governs the flow of electric current; measured in volts

Rechargeable battery – battery which can be used and recharged multiple times, though multiple 'cycles'
Resistance – property of circuit or appliance which resists the flow of electric current, resulting on some electrical energy being converted to heat energy; measured in ohms
Sealed lead-acid (**SLA**) – type of rechargeable battery
Self-discharge – term used to explain the fact that batteries slowly discharge when not in use
Short circuit current (**Isc**) – current through a solar cell when the voltage is zero
Social enterprise – organisation that applies commercial strategies to maximise social and environmental improvements
Solar cell – a semiconductor electrical device that converts the light energy into electricity
Solar lantern – lantern which contains a rechargeable battery and incorporates or is powered by a separate solar module
Solar module – groups of solar cells wired together and encapsulated to generate electrical power
Solar constant – amount of radiation arriving at the edge of the earth's atmosphere from the sun
Solar home system (**SHS**) – an off-grid solar system used to provide power for households – usually for lights and small appliances; solar-home-systems typically use modules between 10 Wp and 100 Wp rating and include a charge controller and rechargeable sealed lead-acid battery
Solar incident angle – angle at which sunlight hits a surface
Standard test conditions (**STC**) – accepted set of common test conditions used by manufacturers to test and compare solar modules; the conditions are: 1000 W/m^2 solar irradiance at 25°C and Air Mass 1.5
State of charge (**SoC**) – the amount of charge in a battery, expressed as a percentage of the total battery capacity
Thin film – type of solar PV cell and module made by depositing thin layers of photovoltaic material on a substrate
Tracking – changing the position of a solar module during the day so that it follows and faces the sun as it moves across the sky
USB – abbreviation referring to an increasingly common connecting plug and socket which operates at 5 volts
Volt (**V**) – unit used to measure electric potential at a given point in an electric circuit; often referred to 'the force' which causes electrons to flow
Watt (**W**) – unit measurement of power; one amp multiplied by one volt = one watt
Watt-hour (**Wh**) – energy measure, calculated by multiplying power (watts) by hours
Watt-peak (**Wp**) – maximum power output generated by a PV module under standard test conditions

Index

Notes
(1) The word order is letter by letter;
(2) Locators to photographs, plans and tables are in *italics*; and
(3) Locators in **bold** refer to entries in the glossary.